Hungarian Problem
Book IV

Library of Congress Catalog Card Number 2010940176
Print edition ISBN 978-0-88385-831-8
Electronic edition ISBN 978-0-88385-913-0
Printed in the United States of America
Current Printing (last digit):
10 9 8 7 6 5 4 3 2 1

Hungarian Problem Book IV

Translated and Edited by

Robert Barrington Leigh

and

Andy Liu

University of Alberta

Published and Distributed by

The Mathematical Association of America

MAA PROBLEM BOOKS SERIES

Problem Books is a series of the Mathematical Association of America consisting of collections of problems and solutions from annual mathematical competitions; compilations of problems (including unsolved problems) specific to particular branches of mathematics; books on the art and practice of problem solving, etc.

Aha! Solutions, Martin Erickson

The Alberta High School Math Competitions 1957–2006: A Canadian Problem Book, compiled and edited by Andy Liu

The Contest Problem Book VII: American Mathematics Competitions, 1995–2000 Contests, compiled and augmented by Harold B. Reiter

The Contest Problem Book VIII: American Mathematics Competitions (AMC 10), 2000–2007, compiled and edited by J. Douglas Faires & David Wells

The Contest Problem Book IX: American Mathematics Competitions (AMC 12), 2000–2007, compiled and edited by David Wells & J. Douglas Faires

First Steps for Math Olympians: Using the American Mathematics Competitions, by J. Douglas Faires

A Friendly Mathematics Competition: 35 Years of Teamwork in Indiana, edited by Rick Gillman

Hungarian Problem Book IV, translated and edited by Robert Barrington Leigh and Andy Liu

The Inquisitive Problem Solver, Paul Vaderlind, Richard K. Guy, and Loren C. Larson

International Mathematical Olympiads 1986–1999, Marcin E. Kuczma

Mathematical Olympiads 1998–1999: Problems and Solutions From Around the World, edited by Titu Andreescu and Zuming Feng

Mathematical Olympiads 1999–2000: Problems and Solutions From Around the World, edited by Titu Andreescu and Zuming Feng

Mathematical Olympiads 2000–2001: Problems and Solutions From Around the World, edited by Titu Andreescu, Zuming Feng, and George Lee, Jr.

Problems from Murray Klamkin: The Canadian Collection, edited by Andy Liu and Bruce Shawyer

The William Lowell Putnam Mathematical Competition Problems and Solutions: 1938–1964, A. M. Gleason, R. E. Greenwood, L. M. Kelly

The William Lowell Putnam Mathematical Competition Problems and Solutions: 1965–1984, Gerald L. Alexanderson, Leonard F. Klosinski, and Loren C. Larson

The William Lowell Putnam Mathematical Competition 1985–2000: Problems, Solutions, and Commentary, Kiran S. Kedlaya, Bjorn Poonen, Ravi Vakil

USA and International Mathematical Olympiads 2000, edited by Titu Andreescu and Zuming Feng

USA and International Mathematical Olympiads 2001, edited by Titu Andreescu and Zuming Feng

USA and International Mathematical Olympiads 2002, edited by Titu Andreescu and Zuming Feng

USA and International Mathematical Olympiads 2003, edited by Titu Andreescu and Zuming Feng

USA and International Mathematical Olympiads 2004, edited by Titu Andreescu, Zuming Feng, and Po-Shen Loh

MAA Service Center
P. O. Box 91112
Washington, DC 20090-1112
1-800-331-1622 fax: 1-301-206-9789

Foreword

George Berzsenyi

The appearance of the present volume is a truly important event in the world of mathematics. It is a huge step forward for its Hilbert Prize winning author, since he had to overcome the tragedy of the death of his young protégée and co-author, and had to find the strength to complete the work alone. It is important to those who are involved with the organization of mathematical competitions, since they now have more complete access to the problems of the famous Kürschák Mathematical Competition for the years 1947–1963, and the solutions of those problems. Moreover, it is important to those who are engaged in the teaching and/or learning of creative mathematical problem solving, since Andy Liu's *Hungarian Problem Book IV* is a wonderful vehicle for mastering the process of problem solving in the spirit of the late George Pólya, who was also a product of the Hungarian school of mathematics.

Indeed, Professor Liu's rendition of the problems and solutions of the 51 problems covered in the second half of János Surányi's original Part II of *Matematikai Versenytételek* is much more than a translation. He groups the problems according to subject matter, develops the background necessary for solving the problems in those areas, and then systematically presents the various solutions, adding several new ones as he proceeds. At the end, he also exemplifies the fourth step of Pólya's process, and reflecting on the solutions of some of the problems, arrives at yet other solutions for some of the other problems. Fortunately, the problems are varied and deep enough to allow for such a treatment at the hands of an expert.

In his foreword to Dr. Liu's *Hungarian Problem Book III*, József Pelikán, the distinguished leader of Hungary's teams to the International Mathematics Olympiads, described the careful process by which the problems are chosen. In the present foreword, I want to reflect on the wisdom of the founding fathers of the Kürschák Mathematical Competition in creating a most admirable framework for that competition. Apart from minor adjustments, it has withstood the tests of time for the last 115 years.

By fixing the number of problems at 3 for each competition, one can provide enough variety, both in topic area and in methodology, to have the very best competitors surface. Of course, all 3 problems must be of the right caliber. Hence the problems committee is always most carefully chosen. Since 1894, some of the best mathematicians of Hungary serve on it. The allowance of 4 hours for the solution of the problems turned out to be also ideal. It is not too short to unduly rush anyone and long enough to complete one's work, often with generalizations, alternate solutions, etc. But the most ingenious idea was to allow the free use of books and notes taken along by the contestants. Thus, the problems tested not the factual knowledge of the familiar materials, but the students' ability to think creatively, their ingenuity and know-how. Hence, it is not surprising that so many of the winners of the Kürschák Mathematical Competition became outstanding scientists in their chosen fields. While in the beginning, only the first and second prize winners were recognized, later, when the number of contestants grew to several hundred, the original format allowed for the identification of several more outstanding students. There are also some years when even the best contestant doesn't get first prize, while in other years several share the first or second prize — it all depends on the caliber of the work turned in.

As far as I know, no other competition reflects the strength of the field of contestants so realistically. At this point I should also add that the organizers of the competition make a huge effort to provide proper recognition in well-publicized ceremonies not only to the winners, but to their teachers too — yet another trait worthy to be emulated.

With respect to the birth of the Kürschák Mathematical Competition, it is instructive to reflect first on the special circumstances which made it reasonable for such a program to be conceived in the minds of the scientists of Hungary back in the 1890s. To this end, I will briefly review the history of Hungary.

Originating somewhere in the Northern parts of present-day China, the Hungarians arrived at the Carpathian Basin around 895 A.D., and established a Christian kingdom there in the Year 1000. Later they withstood the devastation of the Mongolians (13th Century), were partially conquered by the Turks (16th–17th Century), and became more and more subjugated by the Austrians as part of the Habsburg Empire. Following several unsuccessful revolutions, they finally managed to arrive at a compromise with Austria in 1867, which restored most of the territory and some of the earlier independence of the Kingdom of Hungary. At that time, Hungary was nearly $3\frac{1}{2}$ times its present size, with a population of over 20 million, and rich in minerals, forests, and land for agricultural use. The future was promising, and the country was feverishly preparing for its upcoming Millennium, the 1000th anniversary of The Conquest. Hence, it was not unreasonable to think big on that historic occasion, for Hungary was a large country at that time.

In a huge spurt of effort, hundreds of public buildings were erected across the country, including schools, hospitals, banks, museums, etc. In Budapest in 1895 alone, 595 new apartment houses were built, containing nearly 13,000 rooms. It is also notable that 400 new schools were built throughout the country for the millennial year. The world-famous Parliament was completed at that time too, as well as the subway, which was the first in Continental Europe, and three more bridges across the River Danube. This period also saw the construction of several public buildings of later fame, like the Opera House, the Fishermen's Bastion, the Heroes' Square, the Southern and Western Railway Stations, the Museum of Fine Arts, the Main Customs House, and the Art Gallery. In fact, the entire city was redesigned, with three rings of boulevards leading to the bridges and an avenue leading from the city center (of Pest) to the City Park, which was the main venue for the celebratory events. A year-long World Exhibition was organized there, where a proud nation displayed its pursuits in the name of progress and peace.

Thus, it was not surprising that some of the intellectual leaders of the country were thinking on a larger scale too. The groundwork for the latter was laid by the physicist Baron Loránd Eötvös, who was the founder and first president of the Hungarian Physical and Mathematical Society in 1891. He said to his colleagues: "We have to raise the flag of science so high that it should be visible beyond our borders". In that spirit, upon his appointment as Minister of Culture and Education in 1894, members of that Society initiated a "Student Competition". It was named after Eötvös following his death in 1919, and renamed after József Kürschák in 1947 when the physicists wanted to have their competition bear the name of Eötvös.

Kürschák was a professor of mathematics and a strong proponent of this competition; it was a revised edition of his compilation of the competition materials covering the years 1894–1928 that was translated as *Hungarian Problem Books I & II* in 1960. Following Andy Liu's wonderful *Hungarian Problem Books III & IV*, one might ask: What about the problems of the Kürschák Mathematical Competition since 1963? It turns out that the late Professor János Surányi published Parts III and IV of the compilation in Hungarian; they cover the years 1964–1997. Reports on the last 12 years can be found in *KöMaL*, which is the accepted abbreviation for *Középiskolai*

Matematikai és Fizikai Lapok, Hungary's famous high school mathematics and physics journal, which also dates back to 1893. They provide enough first class material for three more Hungarian problem books.

In closing and in the tradition of the previous forewords to the *Hungarian Problem Books*, I should comment on the accomplishments of the 1947–1963 winners of the Kürschák Mathematical Competitions. Instead, I hereby recommend that we give them a bit more time to complete their scientific careers. As one can learn from perusing the internet, many of them are still deeply involved in their respective fields. I am certain that many of them will surface as the best in their disciplines just like the earlier winners of the Kürschák Mathematics Competitions.

Instead, I will close this foreword by expressing my appreciation to my friend, Andy, for this opportunity to write about Hungary's famous competition, which served as a model for all modern mathematical competitions, including the olympiads, throughout the world. And I want to thank him for transplanting this wonderful treasure from the less familiar Hungarian into the English language.

Denver, 2010

Contents

Preface

The present book is a continuation of *Hungarian Problem Book III*, #42 in the Anneli Lax New Mathematical Library Series. It is the translation of the second half of Volume Two of the original Hungarian work. It covers the years from 1947 to 1963, except that the competition was not held in 1956 because of political events, as was the case from 1944 to 1946. After World War II, the competition was renamed after the mathematician József Kürschák instead of after the physicist Loránd Eötvös.

The underlying philosophy of this book is the same as its predecessor. We assume that the reader is familiar with Book III, and will not duplicate the discussions conducted there. The two books should be used in conjunction.

The present book consists of four chapters. In Chapter 1, the contest problems are given in chronological order. There are forty-eight problems altogether. They are classified by subject into twelve sets. Within each set, the four problems are listed in ascending order of estimated difficulty. A Problem Index facilitates the location of solutions to individual problems.

In Chapter 2, the Theorems in *Hungary Problem Book III* are restated, and additional theorems are provided with proofs. In Chapter 3, the solutions to the problems are given set by set.

In Book III, we referred to Pólya's four-step method in problem-solving, focusing primarily on the first three steps, namely, understanding the problem, making a plan, and carrying out the plan. In this book, we will focus on the fourth step, looking back. This is carried out in Chapter 4.

The high school helpers I engaged when I worked on Book III have since moved on in various directions. For the present book, I am fortunate to have been able to recruit as co-author **Robert Barrington Leigh**. Like my earlier helpers, he was also a member of my Mathematics Club. Robert represented Canada twice in the International Mathematical Olympiad. During his undergraduate years at the University of Toronto, he gained much recognition for his outstanding performance in the prestigious William Lowell Putnam Mathematics Competition of the Mathematical Association of America.

Unfortunately, Robert passed away at age twenty in August, 2006, even before completing his undergraduate degree. This book is dedicated to his memory. His contribution, mainly in clever alternative solutions, is significant. I would like to highlight the second solution to Problem 1949.3 in Problem Set 5. Robert found that solution, along with his friend Richard Travis Ng, when they were in grades 6 and 7 respectively. The result was published in Mathematics Competitions **10** (1997) 38–43 under the title "Zigzag", and later republished in the Hungarian journal Abacus **4** (1998) 318–320 under the title "Cikcakk".

I want to thank Don Albers and the editorial committee of the Mathematical Association of America for encouragement and support, and Beverly Ruedi for editorial guidance. I want to thank my Hungarian-American friend George Berzsenyi for writing the Foreword. I would like again to pay tribute to the unidentified Hungarian mathematicians who proposed such inspiring problems in the first place.

Edmonton, 2010

List of Winners

1947	Pollák György, Ungár Péter (IV)
1948	Czipszer János (IV), Neuwirth Sándor (IV)
1949	Korányi Ádám (IV), Róna Péter
1950	Bognár János, Szekerka Pál (III)
1951	Szekerka Pál (IV), Kálmán Lajos (IV)
1952	Kálmán Lajos, Kántor Sándor (IV), Horváth Ákos, Nagy Tibor
1953	Surányi Péter, Vigassy József (IV)
1954	Vigassy József, Csiszár Imre (III), Kovács László
1955	Krammer Gergely
1957	Makkai Mihály
1958	Kisvölcsey Jenő (IV), Bollobás Béla (II)
1959	Jelitai Árpád, Halász Gábor, Muszélyi György, Arató Péter, Csanak György
1960	Mezei Ferenc, Bollobás Béla (IV)
1961	Kóta József (IV), Bollobás Béla, Molnár Emil
1962	Máté Attila (II), Kóta József
1963	Máté Attila, Gerencsér László (IV)

Remark

Names in Hungary are given with surnames preceding given names. However, apart from this list, we follow the western convention. Students normally take the contest shortly after graduation from high school. Those winners whose names are followed by Roman numerals were winners while still in high school, in year II, III or IV. After being a winner in 1962 as a Year II high school student, Attila Máté skipped the final two years of high school. He was a winner again in 1963 as a recent graduate.

1

Kürschák Mathematics Competition Problems

1947

Problem 1. Prove that if n is a positive odd integer, then $46^n + 296 \cdot 13^n$ is divisible by 1947. (Solution is on p. 32.)

Problem 2. Prove that in any group of six people, either there are three people who know one another or three people who do not know one another. Assume that "knowing" is a symmetric relation. (Solution is on p. 26.)

Problem 3. The radius of each small disc is half that of the large disc. How many small discs are needed to cover the large disc completely? (Solution is on p. 63.)

1948

Problem 1. It was Saturday on the 23rd October, 1948. Can one conclude that New Year falls more often on Sundays than on Mondays? (Solution is on p. 29.)

Problem 2. Prove that except for any tetrahedron, no convex polyhedron has the property that every two vertices are connected by an edge. Degenerate polyhedra are not considered. (Solution is on p. 74.)

Problem 3. Prove that from any set of n positive integers, a non-empty subset can be chosen such that the sum of the numbers in the subset is divisible by n. The subset may be equal to the whole set. (Solution is on p. 34.)

1949

Problem 1. Prove that $\sin A + \frac{1}{2}\sin 2A + \frac{1}{3}\sin 3A > 0$ if $0^\circ < A < 180^\circ$. (Solution is on p. 65.)

Problem 2. Let P be any point on the base of a given isosceles triangle. Let Q and R be the intersections of the equal sides with lines drawn through P parallel to these sides. Prove that the reflection of P about the line QR lies on the circumcircle of the given triangle. (Solution is on p. 45.)

Problem 3. Which positive integers cannot be expressed as sums of two or more consecutive positive integers? (Solution is on p. 34.)

1950

Problem 1. On a certain day, a number of readers visited a library. Each went only once. Among any three readers, two of them met at the library on that day. Prove that there were two particular instants such that each reader was in the library at one of the two instants. (Solution is on p. 23.)

Problem 2. Three circles k_1, k_2 and k_3 on a plane are mutually tangent at three distinct points. The point of tangency of k_1 and k_2 is joined to the other two points of tangency. Prove that these two segments or their extensions intersect k_3 at the endpoints of one of its diameters. (Solution is on p. 50.)

Problem 3. Let a_1, b_1, c_1, a_2, b_2 and c_2 be real numbers such that for any integers x and y, at least one of $a_1x + b_1y + c_1$ and $a_2x + b_2y + c_2$ is an even integer. Prove that either all of a_1, b_1 and c_1 are integers or all of a_2, b_2 and c_2 are integers. (Solution is on p. 39.)

1951

Problem 1. $ABCD$ is a square of side a. E is a point on BC with $BE = \frac{a}{3}$. F is a point on DC extended with $CF = \frac{a}{2}$. Prove that the point of intersection of AE and BF lies on the circumcircle of $ABCD$. (Solution is on p. 44.)

Problem 2. For which positive integers m is $(m-1)!$ divisible by m? (Solution is on p. 33.)

Problem 3. A plane can be covered completely by four half-planes. Prove that three of these four half-planes are sufficient for covering the plane completely. (Solution is on p. 64.)

1952

Problem 1. The centers of three mutually disjoint circles lie on a line. Prove that if a fourth circle is tangent to each of them, then its radius is at least the sum of the radii of the other three circles. (Solution is on p. 51.)

Problem 2. Let n be an integer greater than 1. From the integers from 1 to $3n$, $n+2$ of them are chosen arbitrarily. Prove that among the chosen numbers, there exist two of them whose difference is strictly between n and $2n$. (Solution is on p. 36.)

Problem 3. Let $\frac{1}{2} < k < 1$. Let A', B' and C' be points on the sides BC, CA and AB, respectively, of the triangle ABC such that $BA' = kBC$, $CB' = kCA$ and $AC' = kAB$. Prove that the length of the perimeter of the triangle $A'B'C'$ does not exceed k times the length of the perimeter of ABC. (Solution is on p. 61.)

1953

Problem 1. Let n be an integer greater than 2. Two subsets of $\{1, 2, \ldots, n-1\}$ are chosen arbitrarily. Prove that if the total number of elements in the two sets is at least n, then there is one element from each subset such that their sum is n. (Solution is on p. 36.)

Problem 2. Let n be a positive integer and let d be a positive divisor of $2n^2$. Prove that $n^2 + d$ is not a perfect square. (Solution is on p. 29.)

Problem 3. In the convex hexagon $ABCDEF$, the sum of the interior angles at A, C and E is equal to the sum of the interior angles at B, D and F. Prove that opposite angles of the hexagon are equal. (Solution is on p. 68.)

1954

Problem 1. In a convex quadrilateral $ABCD$, $AB + BD \le AC + CD$. Prove that $AB \le AC$. (Solution is on p. 58.)

Problem 2. Prove that if every planar section of a three-dimensional solid is a circle, then the solid is a sphere. (Solution is on p. 76.)

Problem 3. Prove that in a round-robin tournament without ties, there must be a contestant who will list all of his opponents when he lists the ones whom he beats as well as the ones beaten by those whom he beats. (Solution is on p. 22.)

1955

Problem 1. In the quadrilateral $ABCD$, AB is parallel to DC. Prove that if $\angle BAD < \angle ABC$, then $AC > BD$. (Solution is on p. 59.)

Problem 2. How many five-digit multiples of 3 contains the digit 6? (Solution is on p. 30.)

Problem 3. The three vertices of a certain triangle are lattice points. There are no other lattice points on its perimeter but there is exactly one lattice point in its interior. Prove that this lattice point is the centroid of the triangle. (Solution is on p. 48.)

1957

Problem 1. Let ABC be an acute triangle. Consider the set of all tetrahedra with ABC as base such that all lateral faces are acute triangles. Find the locus of the projection onto the plane of ABC of the vertex of the tetrahedron which ranges over the above set. (Solution is on p. 75.)

Problem 2. A factory manufactures several kinds of cloth, using for each of them exactly two of six different colors of silk. Each color appears on at least three kinds of cloth, each with a distinct second color. Prove that there exist three kinds of cloth such that between them, all six colors are represented. (Solution is on p. 24.)

Problem 3. Let n be a positive integer and $\langle a_1, a_2, \ldots, a_n, \rangle$ be any permutation of $1, 2, \ldots, n$. Determine the maximum value of the expression $|a_1 - 1| + |a_2 - 2| + \cdots + |a_n - n|$. (Solution is on p. 38.)

1958

Problem 1. Six points are given on the plane, no three collinear. Prove that three of these points determine a triangle with an interior angle not less than $120°$. (Solution is on p. 64.)

Problem 2. Let u and v be integers such that $u^2 + uv + v^2$ is divisible by 9. Prove that each of u and v is divisible by 3. (Solution is on p. 33.)

Problem 3. $ABCDEF$ is a convex hexagon in which opposite edges are parallel. Prove that triangles ACE and BDF have equal area. (Solution is on p. 42.)

1959

Problem 1. Prove that

$$\frac{x^n}{(x-y)(x-z)} + \frac{y^n}{(y-z)(y-x)} + \frac{z^n}{(z-x)(z-y)}$$

is an integer, where x, y and z are distinct integers and n is a nonnegative integer. (Solution is on p. 40.)

Problem 2. A vertical pole stands on a horizontal plane. The distances from the base of the pole to three other points on the plane are 100, 200 and 300 meters respectively. The sum of the angles of elevation from these three points to the top of the pole is $90°$. What is the height of the pole? (Solution is on p. 72.)

Problem 3. On a certain day, three men visited a friend who is hospitalized. On the same day, their wives did likewise. None of the six visitors went to the sickroom more than once. Each man met the wives of the other two in the sickroom. Prove that at least one of them met his own wife in the sickroom. (Solution is on p. 23.)

1960

Problem 1. Among any four members of a group of travelers, there is one who knows all the other three. Prove that among each foursome, there is one who knows all of the other travelers. Assume that “knowing” is a symmetric relation. (Solution is on p. 27.)

Problem 2. In the infinite sequence $a_1, a_2, a_3, \ldots$ of positive integers, $a_1 = 1$ and $a_k \leq 1 + a_1 + a_2 + \cdots + a_{k-1}$ for $k > 1$. Prove that every positive integer either appears in this sequence or can be expressed as the sum of distinct members of the sequence. (Solution is on p. 35.)

Problem 3. In the square $ABCD$, E is the midpoint of AB, F is on the side BC and G on CD such that AG and EF are parallel. Prove that FG is tangent to the circle inscribed in the square. (Solution is on p. 54.)

1961

Problem 1. Consider the six distances determined by four points in a plane. Prove that the ratio of the largest of these distances to the smallest cannot be less than $\sqrt{2}$. (Solution is on p. 61.)

Problem 2. Prove that if a, b and c are positive real numbers, each less than 1, then the products $(1-a)b$, $(1-b)c$ and $(1-c)a$ cannot all be greater than $\frac{1}{4}$.(Solution is on p. 38.)

Problem 3. Given two circles, exterior to each other, a common inner and a common outer tangents are drawn. The resulting points of tangency define a chord in each circle. Prove that the point of intersection of these two chords or their extensions is collinear with the centers of the circles.(Solution is on p. 57.)

1962

Problem 1. Let n be a positive integer. Consider all ordered pairs (u, v) of positive integers such that the least common multiple of u and v is n. Prove that the number of such pairs is equal to the number of positive divisors of n^2. (Solution is on p. 31.)

Problem 2. Prove that it is impossible to choose more than n diagonals of a convex polygon with n sides, such that every pair of them having a common point. (Solution is on p. 25.)

Problem 3. Let P be a point in or on a tetrahedron $ABCD$ which does not coincide with D. Prove that at least one of the distances PA, PB and PC is shorter than at least one of the distances DA, DB and DC. (Solution is on p. 74.)

1963

Problem 1. Let p and q be integers greater than 1. There are pq chairs, arranged in p rows and q columns. Each chair is occupied by a student of different height. A teacher chooses the shortest student in each row; among these the tallest one is of height a. The teacher then chooses the tallest student in each column; among these the shortest one is of height b. Determine which of $a < b, a = b$ and $a > b$ are possible by an appropriate rearrangement of the seating of the students. (Solution is on p. 21.)

Problem 2. Prove that $(1 + \sec A)(1 + \csc A) > 5$ if A is an acute angle. (Solution is on p. 66.)

Problem 3. Prove that if a triangle is not obtuse, then the sum of the lengths of its medians is not less than or equal to four times the radius of the triangle's circumcircle. (Solution is on p. 60.)

Problem Index

Year	Problem 1	Problem 2	Problem 3
1947	Set 4	Set 2	Set 10
1948	Set 3	Set 12	Set 4
1949	Set 11	Set 7	Set 5
1950	Set 1	Set 8	Set 6
1951	Set 7	Set 4	Set 10
1952	Set 8	Set 5	Set 9
1953	Set 5	Set 3	Set 11
1954	Set 9	Set 12	Set 1
1955	Set 9	Set 3	Set 7
1957	Set 12	Set 2	Set 6
1958	Set 10	Set 4	Set 7
1959	Set 6	Set 11	Set 1
1960	Set 2	Set 5	Set 8
1961	Set 10	Set 6	Set 8
1962	Set 3	Set 2	Set 12
1963	Set 1	Set 11	Set 9

2

Background

In Hungarian Problem Book III, we covered a large number of theorems in basic mathematics. Not much else is needed to tackle the problems in the current volume. We list them again without further discussion, and add two sections on theorems not covered in the earlier volume.

2.1 Theorems in Combinatorics

Principle of Mathematical Induction *If S is a set of positive integers such that $1 \in S$, and $n+1 \in S$ whenever $n \in S$, then S is the set of all positive integers.*

Well Ordering Principle *Any non-empty set of positive integers has a minimum.*

Extremal Value Principle *Every non-empty finite set of real numbers has a maximum and a minimum.*

Mean Value Principle *In every non-empty finite set of real numbers, there is at least one which is not less than the arithmetic mean of the set, and at least one not greater.*

Pigeonhole Principle *Several pigeons are stuffed into several holes. If there are more pigeons than holes, then at least one hole contains at least two pigeons. If there are more holes than pigeons, then there is at least one empty hole.*

Finite Union Principle *The union of finitely many finite sets is also finite.*

Parity Principle *The sum of two even integers is even, the sum of two odd integers is also even, while the sum of an odd and an even integer is odd. Moreover, an odd integer can never be equal to an even integer.*

Multiplication Principle $|A \times B| = |A| \cdot |B|$.

Addition Principle $|A \cup B| = |A| + |B|$ *if A and B have no elements in common.*

Pascal's Formula *For* $1 \le k \le n-1$, $\binom{n}{k} = \binom{n-1}{k-1} + \binom{n-1}{k}$.

Binomial Theorem .

$$(1+x)^n = \binom{n}{0} + \binom{n}{1}x + \cdots + \binom{n}{n}x^n.$$

Graph Parity Theorem *The sum of degrees of all vertices in a graph is equal to twice the number of edges of the graph.*

Cycle Decomposition Theorem *A regular graph of degree 2 is a union of disjoint cycles.*

Bipartite Cycle Theorem *Every cycle in a bipartite graph has even length.*

2.2 Additional Theorems in Combinatorics

In Graph Theory, a **tree** is defined as a connected graph without cycles. Of the three graphs in the diagram below, each with four vertices, the first one is not connected and the last one has a cycle. Only the middle one is a tree.

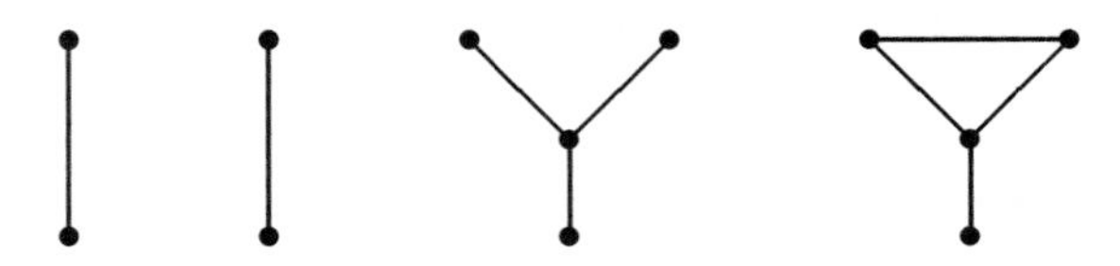

Tree Formula *Let V and E be the numbers of vertices and edges, respectively, of a tree. Then $E = V - 1$.*

Proof Consider a physical model of a tree, where the vertices are beads and edges are threads connecting the beads. Pick up one bead and pull. Since a tree is connected, the whole graph comes off the table. Since a tree has no cycles, every bead except the one being held is dangling at the lower end of a different thread. This establishes a one-to-one correspondence between the beads and the threads apart from the one bead being held. It follows that $E = V - 1$. ■

A tree is an example of a **planar graph**. When drawn without crossing edges, a planar graph divides the plane into regions called **faces**. The infinite region is also considered one of the faces. Note that the concept of faces is meaningless for non-planar graphs.

Every convex polyhedron may be considered a graph, with the same vertices and edges. Such graph are always planar. This can be seen as follows. Imagine that the polyhedron is drawn on a balloon, with the snout of the balloon in one of the faces.

Pull the snout of the balloon outward without ripping anything, until we have a thin rubber disc. This will be the planar representation of the graph, with the face of the polyhedron which contains the snout becoming the infinite region of the graph.

Euler's Formula *Let V, E and F denote the numbers of vertices, edges and faces of a convex polyhedron. Then $V - E + F = 2$.*

Proof We shall prove a more general result for planar graphs, namely, $V + F = E + C + 1$, where C denotes the number of components or connected pieces of the graph. Let us erase all the edges and then put them back one at a time. When there are no edges, we have $E = 0, V = C$ and $F = 1$. The result is true. Note that V never changes. Suppose the result continues to be true after a number of edges have been restored. Consider when the next edge is restored. Suppose it connects two vertices not so far connected. It will not divide an existing face into two, so that F remains constant. On the other hand, C goes down by 1 since two pieces have been joined into one. It follows that the increase of E by 1 is offset by the decrease of C by 1, so that the result continues to hold. Suppose the newly restored edge connects two vertices already connected. It will divide an existing face into two, so that F goes up by 1. On the other hand, C reamins constant. It follows that the increase of E by 1 is balanced by the increase of F by 1, so that the result holds also in this case. When all the edges have been restored, we have $V + F = E + C + 1$. For a graph derived from a convex polyhedron, we have $C + 1$ so that $V - E + F = 2$. ■

As an application of Euler's Formula, we prove that the complete graph on 5 vertices is non-planar. Suppose to the contrary that it is planar. Then it must satisfy Euler's Formula. Since $V = 5$ and $E = \binom{5}{2} = 10$, we have $F = 7$. Now each of the 7 faces is bounded by at least 3 edges, so that a total of at least 21 boundaries are needed. However, each of the 10 edges can only serve as a boundary to 2 faces, providing exactly 20 boundaries. We have a contradiction.

There is a surprising application of this fact. A region of the plane is said to be convex if for any two points in the region, the line segment joining them also lies in the region. Examples of convex sets are circles, triangles and line segments. A quadrilateral may be convex or non-convex. The diagonals of a quadrilateral intersect each other if and only if the quadrilateral is convex.

The **convex hull** of a set of points is the smallest convex region which contains all of the points in that set. The convex hull of three collinear points is the segment joining two and containing the third. The convex hull of three non-collinear points is the triangle with them as the vertices.

A well-known problem asks for a proof that given five points on the plane, no three collinear, some four of them are the vertices of a convex quadrilateral. The usual proof considers the convex hull of the five points. If it is a convex pentagon or quadrilateral, the result is trivial. Suppose it is a triangle ABC. Then the other two points D and E are inside ABC. The line DE divides ABC into two pieces. By the Pigeonhole Principle, one of them contains two of the three vertices, say B and C. Then B and C will form a convex quadrilateral with D and E.

Here is an alternative proof. Treat the five points as vertices, and join every two of them by straight line segments. Two of the segments must cross, as otherwise the graph would be planar. Now the four endpoints of these two segments form a convex quadrilateral.

We now give another application of Euler's Formula. In Hungarian Problem Book III, we claimed that the Utility Graph is non-planar. This graph is also called the complete bipartite graph with 3 vertices on each side. Suppose to the contrary that it is planar. Then it must satisfy Euler's Formula. Since $V = 6$ and $E = 3 \times 3 = 9$, we have $F = 5$. Now each of the 5 faces is bounded by at least 4 edges. This is because the graph is bipartite. If it contains a triangle, then two vertices of the triangle must be on the same side, by the Pigeonhole Principle. However, vertices on the same side are not joined. It follows that a total of at least 20 boundaries are needed. However, each of the 9 edges can only serve as a boundary to 2 faces, providing exactly 18 boundaries. We have a contradiction.

2.3 Theorems in Number Theory

Distributive Theorem *If $a + b = c$ and d divides two of a, b and c, then it also divides the third.*

Linearity Theorem *The greatest common divisor d of a and b is expressible as a* **linear combination** *of these two numbers.*

Division Algorithm *Suppose d does not divide a. We know from arithmetic that by performing the so-called "long division", we will get a quotient q and a remainder r which satisfies $0 < r < d$, such that $a = dq + r$. This process is known as the Division Algorithm.*

Relatively Prime Divisibility Theorem *If $a|bc$ and a is relatively prime to b, then $a|c$.*

Prime Divisibility Theorem *If a prime p divides the product ab, then $p|a$ or $p|b$.*

Fundamental Theorem of Arithmetic *The decomposition of any positive integer greater than 1 into primes is unique, if the prime factors are to be arranged in non-descending order.*

2.4 Theorems in Algebra

Arithmetic-Geometric Mean Inequality *For any* positive *real numbers $x_1, x_2, \ldots, x_n$, we have*

$$M_1(x_1, x_2, \ldots, x_n) \geq M_0(x_1, x_2, \ldots, x_n).$$

Equality holds if and only if $x_1 = x_2 = \cdots = x_n$.

Power Means Inequality *For any real numbers $x_1, x_2, \ldots, x_n$, we have*

$$\begin{aligned} &M_{-\infty}(x_1, x_2, \ldots, x_n) \\ &\leq M_1(x_1, x_2, \ldots, x_n) \\ &\leq M_{\infty}(x_1, x_2, \ldots, x_n). \end{aligned}$$

Equality holds if and only if $x_1 = x_2 = \cdots = x_n$.

Cauchy's Inequality *Let $a_1, a_2, \ldots a_n, b_1, b_2, \ldots, b_n$ be real numbers. Then*

$$\begin{aligned} &(a_1b_1 + a_2b_2 + \cdots + a_nb_n)^2 \\ &\leq (a_1^2 + a_2^2 + \cdots + a_n^2)(b_1^2 + b_2^2 + \cdots + b_n^2), \end{aligned}$$

with equality if and only if for some constant k, $a_i = kb_i$ for $1 \leq i \leq n$ or $b_i = ka_i$ for $1 \leq i \leq n$.

Rearrangement Inequality *Let* $a_1 \le a_2 \le \cdots \le a_n$ *and* $b_1 \le b_2 \le \cdots \le b_n$ *be real numbers. For any permutation* $\langle c_1, c_2, \ldots, c_n \rangle$ *of* $b_1, b_2, \ldots, b_n$, *we have*

$$\begin{aligned} &a_1 b_n + a_2 b_{n-1} + \cdots + a_n b_1 \\ &\le a_1 c_1 + a_2 c_2 + \cdots + a_n c_n \\ &\le a_1 b_1 + a_2 b_2 + \cdots + a_n b_n. \end{aligned}$$

2.5 Additional Theorems in Algebra

For any real number t, the tth **power mean** of n positive numbers $x_1, x_2, \ldots, x_n$ is defined as

$$M_t = \sqrt[t]{\frac{x_1^t + x_2^t + \cdots + x_n^t}{n}}.$$

For $t = 1$, we have the arithmetic mean

$$M_1 = \frac{x_1 + x_2 + \cdots + x_n}{n}.$$

For $t = 2$, we have the **root-mean square**

$$M_2 = \sqrt{\frac{x_1^2 + x_2^2 + \cdots + x_n^2}{n}}.$$

Root-Mean Square Arithmetic Mean Inequality *For any n positive numbers $x_1, x_2, \ldots, x_n$, $M_1 \le M_2$, with equality if and only if $x_1 = x_2 = \cdots = x_n$.*

Proof The inequality is equivalent to

$$n(x_1^2 + x_2^2 + \cdots + x_n^2) \ge (x_1 + x_2 + \cdots + x_n)^2,$$

which is in turn equivalent to

$$(n-1)(x_1^2 + x_2^2 + \cdots + x_n^2) \ge 2(x_1 x_2 + x_1 x_3 + \cdots + x_{n-1} x_n).$$

Subtracting the right side from the left side, we have

$$(x_1 - x_2)^2 + (x_1 - x_3)^2 + \cdots + (x_{n-1} - x_n)^2 \ge 0.$$

Clearly, equality holds if and only if $x_1 = x_2 = \cdots = x_n$. ■

A function $f(x)$ is said to be **convex** over an interval if a chord joining any two points on the graph of the function within the interval lies on or above the graph itself. For most functions, it is sufficient to know that the midpoint of the chord lies on or above the graph. This may be expressed algebraically as $f(\frac{x_1+x_2}{2}) \geq \frac{f(x_1)+f(x_2)}{2}$. Examples of convex functions are x^2 over all real numbers, $\sec x$ over the interval $(-90°, 90°)$ and $\csc x$ over the interval $(0°, 180°)$.

Jensen's Inequality *Let $f(x)$ be a convex function over an interval and let $x_1, x_2, \ldots, x_n$ be numbers in this interval. Then*

$$f\left(\frac{x_1 + x_2 + \cdots + x_n}{n}\right) \geq \frac{f(x_1) + f(x_2) + \cdots + f(x_n)}{n}.$$

Proof The case $n = 2$ is just the definition of a convex function. Assuming that the result holds for some $n \geq 2$, we claim that it also holds for $2n$. Indeed,

$$\begin{aligned} & f\left(\frac{x_1 + x_2 + \cdots + x_{2n}}{2n}\right) \\ = {} & f\left(\frac{\frac{x_1+x_2+\cdots+x_n}{n} + \frac{x_{n+1}+x_{n+2}+\cdots+x_{2n}}{n}}{2}\right) \\ \geq {} & \frac{f(\frac{x_1+x_2+\cdots+x_n}{n}) + f(\frac{x_{n+1}+x_{n+2}+\cdots+x_{2n}}{n})}{2} \\ \geq {} & \frac{\frac{f(x_1)+f(x_2)+\cdots+f(x_n)}{n} + \frac{f(x_{n+1})+f(x_{n+2})+\cdots+f(x_{2n})}{n}}{2} \\ = {} & \frac{f(x_1) + f(x_2) + \cdots + f(x_{2n})}{2n}. \end{aligned}$$

The proof will be complete if we can also show that whenever the result holds for some $n > 2$, it also holds for $n - 1$.

We choose $x_n = \frac{x_1+x_2+\cdots x_{n-1}}{n-1}$. Then

$$\begin{aligned} f(x_n) &= f\left(\frac{x_1 + x_2 + \cdots + x_{n-1}}{n-1}\right) \\ &= f\left(\frac{x_1 + x_2 + \cdots + x_n}{n}\right) \\ &\geq \frac{f(x_1) + f(x_2) + \cdots + f(x_n)}{n}. \end{aligned}$$

Hence $(n-1)f(x_n) \geq f(x_1) + f(x_2) + \cdots + f(x_{n-1})$, which is equivalent to the desired result. ■

A function $f(x)$ is said to be **concave** over an interval if a chord joining any two points on the graph of the function within the interval lies on or below the graph itself. For most functions, it is sufficient to know that the midpoint of the chord lies on or below the graph. This may be expressed algebraically as $f(\frac{x_1+x_2}{2}) \le \frac{f(x_1)+f(x_2)}{2}$. Jensen's Inequality then states that for a concave function $f(x)$ over an interval, $f(\frac{x_1+x_2+\cdots+x_n}{n}) \le \frac{f(x_1)+f(x_2)+\cdots+f(x_n)}{n}$. where $x_1, x_2, \ldots, x_n$ are numbers in this interval.

2.6 Theorems in Geometry

Vertically Opposite Angle Theorem *If the lines AB and CD intersect at a point O, then we have $\angle AOC = \angle BOD$.*

Triangle Inequality *The lengths of two sides of a triangle is greater than the length of the third side.*

SAS Postulate *Triangles ABC and DEF are congruent if $AB = DE$, $\angle CAB = \angle FDE$ and $AC = DF$.*

ASA Theorem *Triangles ABC and DEF are congruent if $AB = DE$, $\angle CAB = \angle FDE$ and $\angle ABC = \angle DEF$.*

Isosceles Triangle Theorem *In Triangle ABC, $AB = AC$ if and only if $\angle ABC = \angle ACB$.*

Sss Inequality *$AB + AC > BD + CD$ for any point D inside triangle ABC.*

Angle-Side Inequality *In triangle ABC, $\angle ABC > \angle ACB$ if and only if $AC > AB$.*

SaS Inequality *In triangles ABC and ABD, $AC = AD$. Then $BD > BC$ if and only if $\angle BAD > \angle BAC$.*

SSS Theorem *Triangles ABC and DEF are congruent if $BC = EF$, $CA = FD$ and $AB = DE$.*

Exterior Angle Inequality *Let D be any point on the extension of the side BC of triangle ABC. Then $\angle ACD > \angle ABC$.*

AAS Theorem *Triangles ABC and DEF are congruent if $BC = EF$, $\angle ABC = \angle DEF$ and $\angle CAB = \angle FDE$.*

HSR Theorem *Triangles ABC and DEF are congruent if $AB = DE$, $BC = EF$ and $\angle CAB = 90° = \angle FDE$.*

Parallel Postulate *Parallelism is transitive.*

Playfair's Theorem *Through a point P not on a line ℓ, exactly one line through P is parallel to ℓ.*

Corresponding Angle Theorem *A line EF cuts two others lines AB and CD at G and H respectively. Then $\angle AGE = \angle CHG$ if and only if AB and CD are parallel.*

Alternate Angle Theorem *A line EF cuts two others lines AB and CD at G and H respectively. Then $\angle BGH = \angle CHG$ if and only if AB and CD are parallel.*

Angle Sum Theorem *The sum of the three angles of a triangle is* $360°$.

Exterior Angle Theorem *Let D be any point on the extension of the side BC of triangle ABC. Then $\angle ACD = \angle ABC + \angle CAB$.*

Parallelogram Theorem *$ABCD$ is a parallelogram if and only if one of the following holds:*

(a) $AB = CD$ *and* $AC = BD$;

(b) $\angle ABC = \angle CDA$ *and* $\angle BCD = \angle DAB$;

(c) $AB = CD$ *and* AB *is parallel to* CD;

(d) AC *and* BD *bisect each other.*

Midpoint Theorem *The segment joining the midpoints of two sides of a triangle is parallel to the third side and equal to half its length.*

Median Trisection Theorem *Let AD and BE be medians of triangle ABC, intersecting each other at the point G. Then we have $AG = 2DG$ and $BG = 2EG$.*

Centroid Theorem *The three medians of a triangle are concurrent at its centroid.*

Incenter Theorem *The three angle bisectors of a triangle are concurrent at its incenter.*

Excenter Theorem *The bisector of an angle of a triangle is concurrent with the bisectors of the exterior angles of the other two angles at one of its excenters.*

Circumcenter Theorem *The perpendicular bisectors of the three sides of a triangle are concurrent at its circumcenter.*

Orthocenter Theorem *The three altitudes of a triangle are concurrent at its orthocenter.*

Angle Bisector Theorem *Let D be a point on the side BC of triangle ABC. Then AD is the bisector of $\angle CAB$ if and only if $\frac{BD}{CD} = \frac{AB}{AC}$.*

Exterior Angle Bisector Theorem *Let D be a point on the extension of the side BC of triangle ABC. Then AD is the bisector of the exterior angle $\angle CAB$ if and only if $\frac{BD}{CD} = \frac{AB}{AC}$.*

Pythagoras' Theorem *In triangle ABC, if $\angle BCA = 90°$, then $AB^2 = BC^2 + CA^2$.*

Pythagoras' Inequality *In triangle ABC, if $\angle BCA < 90°$, then $AB^2 < BC^2 + CA^2$; if $\angle BCA > 90°$, then $AB^2 > BC^2 + CA^2$.*

Median Theorem *$AB^2 + AC^2 = 2AD^2 + 2BD^2$ where AD is a median of triangle ABC.*

AA Theorem *Triangles ABC and DEF are similar if $\angle CAB = \angle FDE$ and $\angle ABC = \angle DEF$.*

sss Theorem *Triangles ABC and DEF are similar if we have $\frac{CA}{AB} = \frac{FD}{DE}$ and $\frac{AB}{BC} = \frac{DE}{EF}$.*

sAs Theorem *Triangles ABC and DEF are similar if $\angle CAB = \angle FDE$ and $\frac{CA}{AB} = \frac{FD}{DE}$.*

Intercept Theorem *Suppose a line intersects three parallel lines at the points B, C and D, and another line intersects the same three parallel lines at the points E, F and G, respectively. Then $\frac{BC}{CD} = \frac{EF}{FG}$.*

Euler's Inequality *The circumradius of a triangle is greater than or equal to its inradius.*

Chord-Radius Theorem *If two of the following statements are true, then so is the third one.*

(a) *A line passes through the center of the circle.*

(b) *A line passes through the midpoint of the chord.*

(c) *A line is perpendicular to the chord.*

Tangent-Radius Theorem *A line passes through a point P on a circle is a tangent to the circle if and only if it is perpendicular to the radius at the point P.*

Intersecting Tangent Theorem *Tangents are drawn from a point P outside a circle to the circle at the points S and T. Then $PS = PT$.*

Thales' Theorem *For any arc, the angle subtended at the center is double the angle subtended at the circle.*

Semicircle-Angle Theorem *Let AB be a diameter of a semicircle. A point C lies on the semi-circle if and only if $\angle BCA = 90°$.*

Cyclic-Quadrilateral Alternate-Angle Theorem *Let $ABCD$ be a convex quadrilateral. Then $ABCD$ is cyclic if and only if $\angle ABD = \angle ACD$.*

Cyclic-Quadrilateral Opposite-Angle Theroem *Let $ABCD$ be a convex quadrilateral. Then $ABCD$ is cyclic if and only if $\angle ABC + \angle CDA = 180°$.*

Tangent-Angle Theorem *Let TP be a tangent to a circle at a point P and A be any other point on the circle. Let B be any point on the circle on the opposite side of AP to T. Then $\angle APT = \angle ABP$.*

Intersecting Chords Theorem *Let $ABCD$ be a convex quadrilateral and P be the point of intersection of AC and BD. Then $ABCD$ is cyclic if and only if $PA \cdot PC = PB \cdot PD$.*

Intersecting Secants Theorem *Let $ABCD$ be a convex quadrilateral and P be the point of intersection of the extension of AB and DC. Then $ABCD$ is cyclic if and only if $PA \cdot PB = PD \cdot PC$.*

Tangent-Secant Theorem *Let A, B and P be points on a circle and let T be the point of intersection of the extension of AB and the tangent to the circle at P. Then $TP^2 = TA \cdot TB$.*

Coordinates-Area Formula *The area of a triangle with coordinates are (x_1, y_1), (x_2, y_2) and (x_3, y_3) for its vertices is given by*

$$\frac{1}{2}|x_1(y_2 - y_3) + x_2(y_3 - y_1) + x_3(y_1 - y_2)|.$$

Pick's Formula *The area of a lattice polygon is given by $I + \frac{1}{2}B - 1$, where I is the number of lattice points inside the polygon and B is the number of lattice points on its boundary.*

Sine-Area Formula *The area of a triangle is given by* $\frac{1}{2}ab\sin\gamma$, *where* γ *is the angle between two sides of lengths* a *and* b.

Law of Sines *Let* α, β *and* γ *be the angles opposite the sides of lengths* a, b *and* c, *respectively, in a triangle. Then*

$$\frac{\sin\alpha}{a} = \frac{\sin\beta}{b} = \frac{\sin\gamma}{c}.$$

Law of Cosines *Let* a, b *and* c *be the side lengths of a triangle, and let* γ *be the angle opposite the side of length* c. *Then*

$$c^2 = a^2 + b^2 - 2ab\cos\gamma.$$

Compound Angle Formulae *For any two angles* α *and* β,

$$\sin(\alpha+\beta) = \sin\alpha\cos\beta + \cos\alpha\sin\beta$$

and

$$\cos(\alpha+\beta) = \cos\alpha\cos\beta - \sin\alpha\sin\beta.$$

Double Angle Formulae *For any angle* α,

$$\sin 2\alpha = 2\sin\alpha\cos\alpha$$

and

$$\cos 2\alpha = \cos^2\alpha - \sin^2\alpha.$$

3

Solutions to Problems

3.1 Problem Set: Combinatorics

Problem 1963.1

Let p and q be integers greater than 1. There are pq chairs, arranged in p rows and q columns. Each chair is occupied by a student of different height. A teacher chooses the shortest student in each row; among these the tallest one is of height a. The teacher then chooses the tallest student in each column; among these the shortest one is of height b. Determine which of $a < b$, $a = b$ and $a > b$ are possible by an appropriate rearrangement of the seating of the students.

First Solution Let A be the tallest among the shortest student in each row, and B be the shortest among the tallest student in each column. Let A be in row i and B be in column j, and let C be the student of height c who is in row i and column j. Note that C may coincide with A or B. We have $a \le c \le b$, so that $a > b$ is impossible. Put the tallest student in each column in the last row. Then the shortest student in this row qualifies as both A and B, yielding $a = b$. Now let her change seats with any other student in the same column. This move does not disqualify her as B, and she is still taller than the shortest student in each row except possibly her new row. We can certainly put a shorter student there, and then $a < b$.

Second Solution We first consider the smallest case with $p = q = 2$. Let the heights of the four students be 1, 2, 3 and 4. Since permutations of rows and columns have no effect on whether $a < b$, $a = b$ or $a > b$, we may assume that the student of height 1 is at the upper left corner. There are six possible configurations.

1	2	1		1	2	1		1	4	1
3	4	3		4	3	3		2	3	2
3	4			4	3			2	4	
1	3	1		1	3	1		1	4	1
2	4	2		4	2	2		3	2	2
2	4			4	3			3	4	

In the first two cases, $a = b = 3$. In the next two cases, $a = b = 2$. In the last two cases, $a = 2 < 3 = b$. It remains to show that $a > b$ is impossible. Suppose to the contrary that there exists such a configuration. Let A be the tallest among the shortest student in each row, and B be the shortest among the tallest student in each column. Then A and B are two different students. If A and B are in the same row, then $a < b$ because of A. If they are in the same column, then $a < b$ because of B. Hence they are not in the same row and not in the same column. Let C be the student in the same row as A and in the same column as B. Let c be the height of C. Then $a < c$ because of A and $c < b$ because of B. Hence $a < b$, a contradiction to our assumption that $a > b$.

Problem 1954.3

Prove that in a round-robin tournament without ties, there must be a contestant who will list all of his opponents when he lists the ones whom he beats as well as the ones beaten by those whom he beats.

First Solution By the Extremal Value Principle, there exists a contestant A who has the highest number of wins. If A does not qualify as the contestant we seek, there must be a contestant B who has beaten A as well as everyone A has beaten. However, B would have a higher number of wins, contradicting the maximality assumption on A.

Second Solution We use mathematical induction on the number n of contestants. For $n = 2$, the winner of the only game of the tournament is the contestant we seek. Suppose the result holds for some $n \geq 2$. Consider now a tournament with $n+1$ contestants. In the sub-tournament omitting one contestant A, one of the n contestants B satisfies the hypothesis of the problem. If A beats B and everyone whom B beats, then A satisfies the hypothesis. Otherwise, B still satisfies the hypothesis.

Third Solution Assemble all contestants in the auditorium. Ask any of them to leave, and take with her everyone she beats. Repeat until nobody is left. We claim that the last one who is asked to leave, either by herself or taking those she beats with her, is the contestant we seek. Obviously, she has beaten everyone asked to leave earlier, or she would have been taken away by one of them. Contestants she does not beat are beaten by one of those asked to leave earlier. This justifies the claim.

Problem 1950.1

On a certain day, a number of readers visited a library. Each went only once. Among any three readers, two of them met at the library on that day. Prove that there were two particular instants such that each reader was in the library at one of the two instants.

First Solution Consider the moment a when the first reader A leaves and the moment b when the last reader B arrives. We may assume that $a < b$ as otherwise every two readers meet. Suppose a reader C is not present at either moment. If C arrives before a, C must be present at a since C leaves after A. Hence C arrives after a. If C leaves after b, C must be present at b since C arrives before B. Hence C leaves before b. It follows that among A, B and C, no two have met. This is a contradiction.

Second Solution Suppose the librarian makes an announcement twice, trying to catch all readers. Clearly, the first moment a should be when the first reader A leaves. If every reader has heard the first announcement, there is nothing else to prove. Otherwise, the second announcement would be made at a later moment b when the first reader B who has not heard the first announcement leaves. Suppose some reader C has not heard either announcement. Then C must have arrived after b, so that among A, B and C, no two have met. This is a contradiction.

Problem 1959.3

On a certain day, three men visited a friend who is hospitalized. On the same day, their wives did likewise. None of the six visitors went to the sickroom more than once. Each man met the wives of the other two in the sickroom. Prove that at least one of them met his own wife in the sickroom.

First Solution Let the men be A, B and C, and let A be the first to arrive. Suppose A is also the last to leave. Since his wife meets B, she must also meet A. Hence we may assume that B is the last to leave. Suppose C does not meet his wife in the sickroom. We may assume by symmetry that C leaves before his wife arrives. Since A meets C's wife, A leaves after C. This means that A is in the sickroom the whole time C is. Since A's wife meets C there, she must also meet A there.

Second Solution Each couple must arrive in the order husband-wife or wife-husband. By the Pigeonhole Principle, two couples arrive in the same order, say A and B before their respective wives. We claim that either A or B meets his wife in the sickroom. Suppose this is not the case. Then A must have left before his wife arrives. Since she meets B, B leaves after A. Now B must also have left before his wife arrives. However, A and B's wife cannot have met, a contradiction.

Third Solution A man meets the other two wives but not his own if and only if his wife is either the first wife to leave or the last wife to arrive. Among the three wives, only one is the first one to leave, and only one is the last to arrive. The third wife must meet her husband.

3.2 Problem Set: Graph Theory

Problem 1957.2

A factory manufactures several kinds of cloth, using for each of them exactly two of six different colors of silk. Each color appears on at least three kinds of cloth, each with a distinct second color. Prove that there exist three kinds of cloth such that between them, all six colors are represented.

First Solution Construct a graph where the vertices R, O, Y, G, B and P represent the colors, and an edge joining two vertices represents a cloth in the colors they represent. Given that every vertex has degree at least 3, we wish to prove that there are three independent edges. Suppose that this is not true. We may assume that RO is an edge. Not both YG and BP can be edges. By symmetry, we may assume that YG is not an edge. Similarly, we may assume that neither is YB. Since Y has degree at least 3, RY, OY and PY are all edges. Then GB is not as otherwise RO, YP and GB form three independent edges. Since G and B also have degrees 3, RG, RB, OG, OB, GP and BP are edges. Now RG, OB and YP form three independent edges.

Second Solution Construct a graph as in the First Solution. Given that every vertex has degree at least 3, we wish to prove that there are three independent edges. We may assume that RO is an edge. Since Y has degree at least 3, we may assume that YG is also an edge. If BP is also an edge, then we have three independent edges. So suppose BP is not an edge. Then at least 3 of RB, OB, YB and GB are edges. By symmetry, we may assume that OB and YB are edges. Now at least 3 of RP, OP, YP and GP are edges. Hence either RP or GP is an edge. The former forms three independent edges with YG and OB, while the latter does so with RO and YB.

Third Solution Construct a graph as in the First Solution. Given that every vertex has degree at least 3, we wish to prove that there are three independent edges. If there is an edge such that its removal does not reduce the degree of any vertex below 3, we will remove it. When all such edges have been removed, we claim that the resulting graph G still have three independent edges. The complement $\overline{G}$ of G is a graph with the same vertices, such that two vertices are joined by an edge if and only if they are not joined by an edge in G. The degree of each vertex in $\overline{G}$ is at most 2. Suppose two of them have degree less than 2. They must be joined by an edge. Otherwise, adding this edge to $\overline{G}$ does not raise any degree above 2, so that this edge should have been deleted from G. It follows that $\overline{G}$ has one of three forms: (1) all vertices have degree 2; (2) one vertex has degree 0 and all others have degree 2; (3) two vertices have degree 1 and are joined by an edge while all others have degree 2. Note that we cannot have three vertices of degree less than 1, since they would have been joined pairwise, and their degrees could not be less than 2. It follows that $\overline{G}$ is one of the graphs shown in the diagram below, and in each case, G has three independent edges shown in dashes.

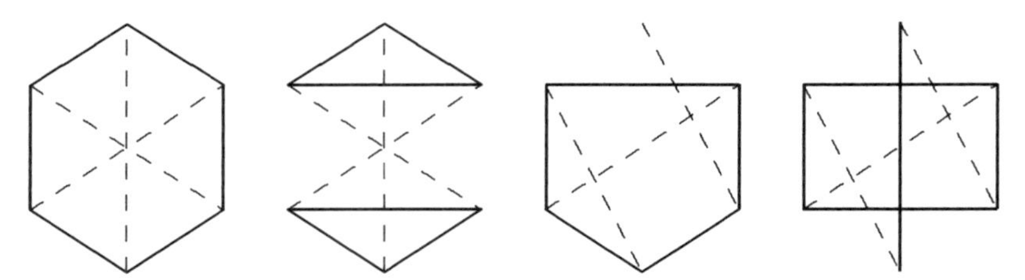

Problem 1962.2

Prove that it is impossible to choose more than n diagonals of a convex polygon with n sides, such that every pair of them have a common point.

First Solution The conclusion of the problem does not depend on the exact shape of the convex polygon. Hence we need only consider the regular one. Divide the diagonals into parallel classes. Then at most one line from each class may be chosen. All lines in the same class has the same perpendicular bisector. If n is odd, this line passes through a vertex and the midpoint of the opposite side. There are exactly n such lines, and hence exactly n classes. If n is even, the perpendicular bisector passes through either two opposite vertices or the midpoints of two opposite sides. Again there are exactly n such lines, and hence exactly n classes. It follows that at most n diagonals can be chosen so that every pair of them have a common point.

Second Solution Suppose we can choose d diagonals such that every pair of them have a common point. Construct a graph where the vertices are the n vertices of the convex polygon and the edges are the d mutually intersecting diagonals. Let a_i denote the number of vertices of degree i. Then $a_1 + a_2 + a_3 + \cdots = n$. Suppose P is a vertex of degree at least 3. Let PQ, PR and PS be three of the chosen diagonals, with PR between the other two. Then R cannot lie on another edge RT as the diagonal RT cannot intersect both PQ and PS. Hence R has degree 1. It follows that a vertex of degree $i > 2$ is joined to $i - 2$ vertices of degree 1. Since each vertex of degree 1 is joined to at most one vertex of degree $i > 2$, we have $a_1 \geq a_3 + 2a_4 + 3a_5 + \cdots$. Now

$$\begin{aligned} 2d &= a_1 + 2a_2 + 3a_3 + \cdots \\ &= (a_1 + 2a_2 + 2a_3 + \cdots) + (a_3 + 2a_4 + 3a_5 + \cdots) \\ &\leq (a_1 + 2a_2 + 2a_3 + \cdots) + a_1 \\ &= 2n. \end{aligned}$$

Third Solution Suppose we can choose d diagonals such that every pair of them have a common point. Construct a graph where the vertices are the n vertices of the convex polygon and the edges are the d mutually intersecting diagonals. Let a_i denote the number of vertices of degree i. Suppose P is a vertex of degree at least 3. Let PQ, PR and PS be three of the chosen diagonals, with PR between the other two. Then R cannot lie on another edge RT as the diagonal RT cannot intersect both PQ and PS. Hence R has degree 1. We now reduce the graph as follows. If there is a vertex of degree 1, remove it and the edge incident with it. Continue until no such vertices remain. Note that some vertices whose initial degree is greater than 1 may end up being removed since its degree may change as other vertices are removed. Suppose we are left with an empty graph. Then we have removed d edges and d vertices. Since the total number of vertices is n, we have $d \leq n$. Suppose there are vertices with positive degree. Then each is of degree 2, since the absence of vertices of degree 1 precludes the existence of vertices of degree $i > 2$. Hence the residual graph is a disjoint union of cycles. We have removed k edges and k vertices, and are left with $d - k$ edges and at least $d - k$ vertices. Hence $d - k \leq n - k$ so that $d \leq n$.

Problem 1947.2

Prove that in any group of six people, either there are three people who know one another or three people who do not know one another. Assume that "knowing" is a symmetric relation.

First Solution Construct a graph with six vertices representing the six people. Two vertices are joined by a blue edge if the two people they represent do not know each other, and by a red edge if they know each other. Consider a particular vertex A and the five edges incident with it. By the Mean Value Principle, at least three of them have the same color, say AB, AC and AD. If any of BC, BD and CD also has the same color, say BC, then we have a triangle ABC whose edges are of the same color. If not, then BCD is a triangle whose edges are of the same color. Such a triangle represent three people who either know one another or do not know one another.

Second Solution Construct a graph with six vertices representing the six people. Two vertices are joined by a blue edge if the two people they represent do not know each other, and by a red edge if they know each other. Define an arrow as two edges meeting at a vertex. The arrow is said to be uniform if both edges have the same color, and mixed otherwise. Consider each of the six vertices. Since it is incident with 5 edges, the number of arrows with this vertex as the middle is $\binom{5}{2} = 10$. If the color-split is 5:0, all arrows there are uniform. If it is 4:1, the number of uniform arrows there is 6. If the split is 3:2, the number is 4. Hence the total number of uniform arrows is at least 24. Now there are $\binom{6}{3} = 20$ triangles in the graph. Each triangle contains 3 arrows. If the three edges of the triangle are of the same color, all 3 arrows are uniform. If not, exactly 1 of them is uniform. Let T be the number of triangles all edges of which are of the same color. Then the total number of uniform arrows is given by

$$3T + (20 - T) = 2T + 20.$$

Since this total is at least 24, we have

$$2T + 20 \geq 24 \quad \text{or} \quad T \geq 2.$$

Problem 1960.1

Among any four members of a group of travelers, there is one who knows all of the other three. Prove that among each foursome, there is one who knows all of the other travelers. Assume that "knowing" is a symmetric relation.

First Solution Construct a graph where the vertices represent the travelers, and two vertices are joined by an edge if and only if the two travelers they represent do not know each other. Let $n \geq 4$ is the number of vertices. We claim that the graph has at least $n - 3$ isolated vertices. We use mathematical induction on n. The case $n = 4$ is just the hypothesis of the problem. Suppose the claim holds for some $n \geq 4$. Consider the next case with $n + 1$ vertices. We first established that there is at least one isolated vertex. Set aside an arbitrary vertex A. By the induction hypothesis, the remaining subgraph has $n - 3$ isolated vertices. Since $n \geq 4$, there is at least one isolated vertex B. If B is not joined to A, then it is an isolated vertex in the whole graph. Suppose B is joined to A. Then no two other vertices are joined to each other. Let C be any other vertex and consider the subgraph obtained when C is set aside. By the induction hypothesis, there is at least one isolated vertex which cannot be either A or B. Since this vertex is not joined to C either, it is an isolated vertex in the whole graph. It follows that we have at least one isolated vertex. Setting it aside and applying the induction hypothesis to the remaining subgraph, it has $n-3$ isolated vertex which must also be isolated in the whole graph since the vertex set aside is itself isolated. Hence there are $n-2$ isolated vertex in the whole graph, completing the inductive argument.

Second Solution Construct a graph where the vertices represent the travelers, and two vertices are joined by an edge if and only if the two travelers they represent do not know each other. This graph has two properties. First, every vertex is of degree at most 2. Suppose to the contrary that vertex A is joined to vertices B, C and D. Then in the foursome A, B, C and D, nobody knows all of the other three. Second, there are no independent edges. Suppose to the contrary that there are two such edges AB and CD. Again, in the foursome A, B, C and D, nobody knows all of the other three. If the graph has no edges, there is nothing to prove. Suppose there is an edge AB. If this is the only edge, then in any foursome, we can choose anyone other than A or B, and this person will know all of the other travelers. It follows that there is a second edge. It must be incident with A or B. By symmetry, we may assume that this edge is AC. If there are no other edges, then in any foursome, we can choose anyone other than A, B and C, and this person will know all of the other travelers. Now a new edge can no longer be incident with A. However, it cannot be independent of AB or AC. Hence it can only be BC. In any foursome, we can choose anyone other than A, B and C, and this person will know all of the other travelers.

3.3 Problem Set: Number Theory

Problem 1948.1

It was Saturday on the 23rd October, 1948. Can one conclude that New Year falls more often on Sundays than on Mondays?

Solution According to the rules of the calendar, a year is a leap year if it is divisible by 4, unless it is also divisible by 100, in which case it is not a leap year unless it is divisible by 400. Hence there are 97 leap years and 303 ordinary years in each cycle. Each ordinary year has 365 days while each leap year has 366 days. When divided by 7, these numbers leave remainders of 1 and 2 respectively. Now $400 + 97 = 497$ is a multiple of 7. Hence the days of the week also repeat itself in a cycle of 400 years. However, since 400 is not a multiple of 7, the days of the week cannot occur with equal frequency on New Year. We now reduce each ordinary year down to its New Year and each leap year down to its New Year preceded by its New Year Eve. Then we have 71 weeks in which the days of the week occur normally. If we can show that there are more Mondays than Sundays among these New Year Eves, then there are more Sundays than Mondays among the New Years. We put on a list the 97 New Year Eves from 1600 to 1996 inclusive. Suppose one of the New Year Eves on the list is a Sunday. Then the next one should be a Friday, since the days of the week have been shifted by $4 + 1 = 5$ days. This holds except for 1696, 1796 and 1896 when 1700, 1800 and 1900 are ordinary years. To compensate for this, we delete from our list the New Year Eves of 1704, 1708, 1804, 1808, 1904 and 1908. From 1696 to the next leap year 1712 on the list, the days of the week have been shifted by $16 + 3 = 19 \equiv 5 \pmod 7$. Moreover, this leaves only 91 New Year Eves on our list.

Hence the days of the week occur with equal frequencies among them. All that remains is to work out the days of the week for the New Year Eves of the six deleted leap years. From October 23 to December 31, the number of days is $8+30+31 = 69 \equiv 6 \pmod 7$. Hence New Year Eve in 1948 is a Friday. From New Eve in 1908 to 1948, we have $40+10 = 50 \equiv 1 \pmod 7$. Hence New Year Eve in 1908 is a Thursday. For 1808, we have $100+24 = 124 \equiv 5 \pmod 7$ so that its New Year Eve is a Saturday. Shifting backwards by 5 days again, New Year Eve in 1708 is a Monday. For 1904, 1804 and 1704, we shift forward by 2 days from 1908, 1808 and 1708, yielding a Saturday, a Monday and a Wednesday, respectively. In summary, in a cycle of 400 years, the numbers of New Years on various days of the week starting from Sunday are 58, 56, 58, 57, 57, 58 and 56. Thus we can conclude that New Year falls more often on Sundays than on Mondays.

Problem 1953.2

Let n be a positive integer and let d be a positive divisor of $2n^2$. Prove that $n^2 + d$ is not a perfect square.

First Solution Suppose $n^2 + d = m^2$ for some positive integer m. Then $m^2 - n^2 = d$ divides $2n^2$. Hence it also divides $2n^2 + 2(m^2 - n^2) = 2m^2$, and therefore it divides the greatest common divisor of $2n^2$ and $2m^2$. Now this greatest common divisor is equal to $2k^2$ where k is the greatest common divisor of m and n. Let $m = ka$ and $n = kb$ for some positive integers a and b. Then $(ka)^2 - (kb)^2$ divides $2k^2$, which means that $a^2 - b^2$ divides 2. Hence $a^2 - b^2 = 1$ or 2. This is a contradiction since the difference of the squares of two positive integers cannot be equal to 1 or 2.

Second Solution We have $2n^2 = kd$ for some positive integer k. Suppose that $n^2 + d = m^2$ for some positive integer m. Then we have $m^2 = n^2 + \frac{2n^2}{k}$ so that $(mk)^2 = n^2(k^2 + 2k)$. By the Fundamental Theorem of Arithmetic, $k^2 + 2k$ must also be the square of a positive integer. However, since we have $k^2 < k^2 + 2k < (k+1)^2$, it falls between two consecutive squares. This contradiction establishes the desired result.

Third Solution We have $2n^2 = kd$ for some positive integer k. Suppose that $n^2 + d = m^2$ for some positive integer m. Then we have $\frac{k+2}{k} = \frac{n^2+d}{n^2} = \frac{m^2}{n^2}$. Now the difference between the numerator and the denominator of $\frac{k+2}{k}$ is 2. When reduced to the lowest terms, this difference either stays at 2 or reduces to 1. On the other hand, the reduced form of $\frac{m^2}{n^2}$ is $\frac{p^2}{q^2}$ where p and q are positive integers with $p > q$. Then $p^2 - q^2 = (p+q)(p-q) \geq p + q \geq 3$. This contradiction establishes the desired result.

Problem 1955.2

How many five-digit multiples of 3 contains the digit 6?

First Solution Every third of the 90000 five-digit numbers is divisible by 3. We now determine how many of these 300000 multiples do not contain the digit 6. The first digit can be chosen in 8 ways, and each of the next three digits may be chosen in 9 ways. If the sum of the other four digits is congruent modulo 3 to 0, we may choose 0, 3 or 9 as the last digit. If the sum is congruent to 1, we may choose 2, 5 or 8. If the sum is congruent to 2, we may choose 1, 4 or 7. Hence there are 17496 such multiples, and the number of multiples with at least one 6 is $30000 - 17496 = 12504$.

Second Solution Of the three integers $10m$, $10m + 1$ and $10m + 2$, either each contains the digit 6 or none does. The same applies to the three integers $10m + 3$, $10m + 4$ and $10m + 5$, and to $10m + 7$, $10m + 8$ and $10m + 9$. Thus numbers that do not contain the digit 6 come in blocks of three consecutive numbers. Now every third number is a multiple of 3, and if we eliminate all numbers not containing the digit 6, it is still true that every third number is a multiple of 3. Of the 90000 five-digit numbers, there are 52488 which do not contain the digit 6. This is because the first digit can be chosen in 8 ways and each of the other digit can be chosen in 9 ways. Of the $90000 - 52488 = 37512$ five-digit numbers containing the digit 6, every third is a multiple of 3. Since 37512 is divisible by 3, the desired number is one third of it or 12504.

Third Solution We consider three cases according to where the digit 6 last appears.

Case 1. The last digit is 6.
Each of the middle three digits may be chosen arbitrarily, in 10 ways. The first digit cannot be 0. If the sum of the other four digits is congruent modulo 3 to 0, we may choose 3, 6 or 9. If the sum is congruent to 1, we may choose 2, 5 or 8. If the sum is congruent to 2, we may choose 1, 4 or 7. Hence there are 3000 such multiples.

Case 2. Only the first digit is 6.
Each of the middle three digits may be chosen arbitrarily, but in only 9 ways since 6 is not permitted. For the last digit, if the sum of the other four digits is congruent modulo 3 to 0, we may choose 0, 3 or 9. If the sum is congruent to 1, we may choose 2, 5 or 8. If the sum is congruent to 2, we may choose 1, 4 or 7. Hence there are 2187 such multiples.

Case 3 The last 6 is one of the middle three digits.
Each digit after it can be chosen in 9 ways, each digit before it except the very first can be chosen in 10 ways, and the very first digit can be chosen in 3 ways. Hence there are 2700 multiples with the last 6 in the fourth place, 2430 multiples with the last 6 in the third place, and 2187 multiples with the last 6 in the second place.

The grand total is $3000 + 2700 + 2430 + 2187 + 2187 = 12504$.

Problem 1962.1

Let n be a positive integer. Consider all ordered pairs (u, v) of positive integers such that the least common multiple of u and v is n. Prove that the number of such pairs is equal to the number of positive divisors of n^2.

First Solution Let the prime factorization of n, u and v be

$$n=\prod_{i=1}^{m} p_i^{k_i},\ u=\prod_{i=1}^{m} p_i^{a_i} \text{ and } v=\prod_{i=1}^{m} p_i^{b_i}.$$

For $1 \le i \le m$, we have $\max\{a_i, b_i\} = k_i$. There are $2k_i + 1$ possible choices for (a_i, b_i), namely,

$$(a_i, b_i) = (0, k_i), (1, k_i), \ldots, (k_i, k_i), (k_i, k_i - 1), \ldots, (k_i, 0).$$

Thus the total number of ordered pairs (u, v) is $(2k_1+1)(2k_2+1)\cdots(2k_m+1)$. On the other hand, the prime factorization of n^2 is $n^2 = \prod_{i=1}^{m} p_i^{2k_i}$. Let $d = \prod_{i=1}^{m} p_i^{c_i}$ be the prime factorization of a divisor d of n^2. For $1 \le i \le m$, there are also $2k_i + 1$ possibilities, namely, $c_i = 0, 1, \ldots, 2k_i$. Hence the total number of divisors of n^2 is $(2k_1+1)(2k_2+1)\cdots(2k_m+1)$, and the desired result follows.

Second Solution We shall establish a one-to-one correspondence between the ordered pairs (u, v) and the divisors of n^2. We first prove a preliminary result. Let w, x, y and z be positive integers such that $\frac{w}{y} = \frac{x}{z}$. Then $\frac{[w,x]}{[y,z]}$ has the same value, where $[a, b]$ denotes the least common multiple of a and b. Now computing $[w, x]$ boils down to finding two positive integers r and s which are as small as possible, such that $wr = xs$ is $[w, x]$. Clearly, $\frac{s}{r}$ is $\frac{w}{x}$ in the simplest form. Since $\frac{w}{x} = \frac{y}{z}$, we also have $[y, z] = yr = zs$. It follows that $\frac{[w,x]}{[y,z]} = \frac{wr}{yr} = \frac{w}{y}$ as desired. Returning to the original problem, let (u, v) be an ordered pair such that $[u, v] = n$. Define $d = \frac{un}{v}$. Since v is a divisor of n, d is a positive integer. Moreover, $\frac{n^2}{d} = \frac{vn}{u}$. Hence d is a divisor of n^2. If $\frac{u_1}{v_1} = \frac{u_2}{v_2}$ where $[u_1, v_1] = [u_2, v_2] = n$, then $\frac{u_1}{u_2} = \frac{v_1}{v_2} = \frac{[u_1,v_1]}{[u_2,v_2]} = \frac{n}{n} = 1$. It follows that $(u_1, v_1) = (u_2, v_2)$, so that each ordered pair is associated with a unique divisor of n^2. From $\frac{u}{d} = \frac{v}{n} = \frac{n}{[d,n]}$, we have $u = \frac{dn}{[d,n]}$ and $v = \frac{n^2}{[d,n]}$. Thus defined, (u, v) is the ordered pair associated with the divisor d of n^2. Since dn and n^2 are both common multiples of d and n, u and v are positive integers. Since $\frac{n}{[d,n]} = \frac{u}{d} = \frac{v}{n} = \frac{[u,v]}{[d,n]}$, we indeed have $[u, v] = n$. This establishes the desired one-to-one correspondence.

3.4 Problem Set: Divisibility

Problem 1947.1

Prove that if n is a positive odd integer, then $46^n + 296 \cdot 13^n$ is divisible by 1947.

First Solution Let $n = 2k + 1$ for some non-negative integer k. Note that $46^2 = 2116 = 1947 + 169$ while $296 \cdot 13 = 3848 = 2 \cdot 1947 - 46$. In congruence modulo 1947,

$$\begin{aligned} 46^n + 296 \cdot 13^n &\equiv 46(46^2)^k + (296 \cdot 13)(13^2)^k \\ &\equiv 46(169)^k - 46(169)^k \\ &\equiv 0. \end{aligned}$$

Second Solution Since n is odd, $n-1$ is even and $x^{n-1} - y^{n-1}$ is divisible by $x^2 - y^2$. For $x = 46$ and $y = 13$, $x^2 - y^2 = 1947$. Now $46^n + 296 \cdot 13^n = 46(46^{n-1} - 13^{n-1}) + (46 + 296 \cdot 13)13^{n-1}$. Since $46 + 296 \cdot 13 = 2 \cdot 1947$, the whole expression is divisible by 1947.

Third Solution Since $1 + 9 + 4 + 7 = 21 \equiv 2 + 1 = 3 \equiv 0 \pmod{3}$, we have $1947 = 3 \times 649$. Since $6 - 4 + 9 = 11 \equiv 0 \pmod{11}$, we have $649 = 11 \times 59$. Thus the prime factorization of 1947 is $3 \times 11 \times 59$. Now $46^n + 296 \cdot 13^n \equiv 1^n + (-1)1^n = 0 \pmod{3}$, $46^n + 296 \cdot 13^n \equiv 2^n + (-1)2^n = 0 \pmod{11}$ and $46^n + 296 \cdot 13^n \equiv (-13)^n + 1 \cdot 13^n = 0 \pmod{59}$. In the last congruence, we have $(-13)^n = -13^n$ since n is odd. It follows that $46^n + 296 \cdot 13^n$ is divisible by 1947.

Problem 1951.2

For which positive integers m is $(m - 1)!$ divisible by m?

Solution For $m = 1$, $0!$ is divisible by 1. For m a prime number, it does not divide any factor in $(m - 1)!$, and hence does not divide $(m - 1)!$. Suppose $m = ab$ where a and b are divisors of m strictly between 1 and m. If $a \neq b$, then both will appear as factors in $(m - 1)!$, and m will divide $(m - 1)!$. If m cannot be expressed as a product of two distinct factors strictly between 1 and m but is nevertheless composite, then it must be the square of a prime number p. Now p^2 divides $(p^2 - 1)!$ if and only if $2p$ appears as a factor in $(p^2 - 1)!$. This is equivalent to $2p < p^2$ or $2 < p$. Indeed, for $p = 2$, $p^2 = 4$ does not divide $(p^2 - 1)! = 3!$. In summary, m divides $(m - 1)!$ if and only if $m = 1$ or m is any composite number other than 4.

Problem 1958.2

Let u and v be integers such that $u^2 + uv + v^2$ is divisible by 9. Prove that each of u and v is divisible by 3.

Solution We have $u^2 + uv + v^2 = (u - v)^2 + 3uv$. Since the left side is divisible by 9, the right side is divisible by 9. Since $3uv$ is divisible by 3, $(u - v)^2$ must be divisible by 3. This means that $u - v$ is divisible by 3, so that $(u - v)^2$ is divisible by 9. Hence $3uv$ is also divisible by 9 or uv divisible by 3. It follows that at least one of u and v is divisible by 3, and since so is $u - v$, both are divisible by 3.

Problem 1948.3

Prove that from any set of n positive integers, a non-empty subset can be chosen such that the sum of the numbers in the subset is divisible by n. The subset may be equal to the whole set.

Solution Let $a_1, a_2, \ldots, a_n$ be the given positive integers. Let us define $b_k = a_1 + a_2 + \cdots + a_k$ for $1 \le k \le n$. Divide the n numbers $b_1, b_2, \ldots, b_n$ by n. If some remainder is 0, we have the desired subset of numbers. If not, then each remainder is one of 1, 2, ..., $n - 1$. By the Pigeonhole Principle, two of the remainders must be the same. Suppose they are obtained when b_i and b_j are divided by n, with $i < j$. Then $a_{i+1} + a_{i+2} + \cdots + a_j = b_j - b_i$ is divisible by n.

3.5 Problem Set: Sums and Differences

Problem 1949.3

Which positive integers cannot be expressed as sums of two or more consecutive positive integers?

First Solution We claim that a number is expressible as a sum of two or more consecutive positive integers if and only if it has an odd divisor greater than 1. Suppose first that $n = (2d + 1)m$ for some positive integers d and m. Then

$$n = (m - d) + (m - d + 1) + \cdots + (m + d).$$

This is a sum of consecutive positive integers unless $m < d$. In this case, we can cancel all negative terms with the corresponding positive terms, leaving behind a sum of consecutive positive integers equal to $(2d + 1)m$. Moreover, this is a sum of at least two terms. Otherwise $-(m - d) = (m + d - 1)$ or $2m = 1$, which is impossible. Suppose now that n is expressible as a sum of two or more consecutive positive integers. Then there exist positive integers a and k such that $n = a + (a + 1) + \cdots + (a + k)$. The sum of the first and the last terms is $2a + k$ while the number of terms is $k + 1$. Since the difference of these two numbers is $2a - 1$, one of them is odd and the even.

Hence $n = \frac{(2a+k)(k+1)}{2}$ is divisible by some odd number greater than 1. This justifies our claim. The only numbers which cannot be expressed as a sum of two or more consecutive positive integers are those which have no odd divisors greater than 1, and these are precisely the powers of 2.

Second Solution A sum of two or more consecutive positive integers may be represented as a staircase, and two copies of the staircase will form a box, by which we mean a rectangle with integer dimensions greater than 1. The diagram below shows two illustrative examples.

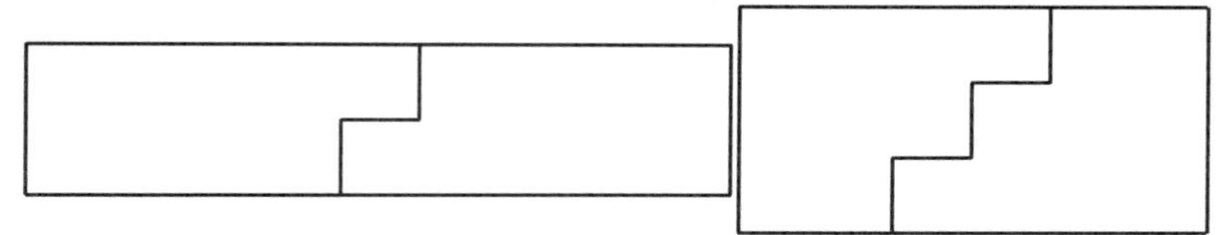

Each box is divided into halves by a zigzag line. Now any such line starts with a vertical segment which we call a zag, followed by a horizontal segment which we call a zig. Then comes another zag followed by a zig, and so on. It ends with a zag. Hence there is an odd number of zigs and zags, which means that there is a middle zig or zag. If we have a middle zig, then the horizontal dimension of the box must be odd. This is because the two staircases are symmetric about this middle zig. Similarly, if we have a middle zag, then the vertical dimension of the box must be odd. A positive integer n is expressible as a sum of two or more consecutive positive integers if we can find a box of area $2n$ with one of its dimensions odd. For such a box to exist, n must have an odd divisor greater than 1. This is the case if n is not a power of 2. Thus all numbers which are not powers of 2 can be expressed as the sum of two or more consecutive positive integers. If n is a power of 2, then so is $2n$. In any box with area $2n$, both dimensions are even. So there will not be a middle column or a middle row. Thus there is nothing that can serve as the middle zig or zag, and it is impossible to divide the box into two identical staircases. Therefore, a power of 2 cannot be expressed as a sum of two or more consecutive integers.

Problem 1960.2

In the infinite sequence $a_1, a_2, a_3, \ldots$ of positive integers, $a_1 = 1$ and $a_k \leq 1 + a_1 + a_2 + \cdots + a_{k-1}$ for $k > 1$. Prove that every positive integer either appears in this sequence or can be expressed as the sum of distinct members of the sequence.

First Solution If n is a positive integer such that $n \leq a_1+a_2+\cdots+a_m$, we shall prove by induction on m that then n is the sum of some of the terms on the right side. This is certainly true for $m = 1$, since $n \leq a_1 = 1$ implies that $n = 1$, and we already have equality. Suppose the result holds for some $m \geq 1$. If $n \leq a_1 + a_2 + \cdots + a_{m-1}$ is still true, we can just apply the induction hypothesis. Hence we may assume that $a_m \leq 1 + a_1 + a_2 + \cdots + a_{m-1} \leq n \leq a_1 + a_2 + \cdots + a_m$. It follows that $0 \leq n - a_m \leq a_1 + a_2 + \cdots + a_{m-1}$. If $n - a_m = 0$, then we simply take $n = a_m$. Otherwise, $n - a_m$ is a positive integer, and by the induction hypothesis, it is the sum of some of $a_1, a_2, \ldots, a_{m-1}$. Hence n is the sum of these along with a_m.

Second Solution If m is the smallest positive integer such that $n \leq a_1 + a_2 + \cdots + a_m$, we shall prove by induction on n that n is the sum of some of the terms on the right side. For $n = 1$, we have $m = 1$ and $n = a_1$. For some fixed $n > 1$, we assume that the result holds for all smaller values. Let m be the smallest positive integer such that $n \leq a + 1 + a_2 + \cdots + a_m$. Define $k = a_1 + a_2 + \cdots + a_m - n$. Then

$$0 \leq k \leq a_m - 1 \leq a_1 + a_2 + \cdots + a_{m-1} < n.$$

If $k = 0$, then $n = a_1 + a_2 + \cdots + a_m$. Otherwise, the induction hypothesis shows that k is the sum of some of $a_1, a_2, \ldots, a_{m-1}$. Hence n is the sum of these along with a_m.

Problem 1953.1

Let n be an integer greater than 2. Two subsets of $\{1, 2, \ldots, n-1\}$ are chosen arbitrarily. Prove that if the total number of elements in the two sets is at least n, then there is one element from each subset such that their sum is n.

Solution Let the two subsets be $A = \{a_1, a_2, \ldots, a_r\}$ and $B = \{b_1, b_2, \ldots, b_s\}$, with $r + s \geq n$. Then $C = \{n - b_1, n - b_2, \ldots, n - b_s\}$ is also an s-element subset of S. Hence there is an element common to A and C. Let it be $a_i = n - b_j$. Then we have $a_i + b_j = n$ as desired.

Problem 1952.2

Let n be an integer greater than 1. From the integers from 1 to $3n$, $n + 2$ of them are chosen arbitrarily. Prove that among the chosen numbers, there exist two of them whose difference is strictly between n and $2n$.

First Solution We may assume that $3n$ is among the chosen numbers. If not, we may simply increase every chosen number by a fixed amount which brings the largest to $3n$. This has no effect on their pairwise differences. If any of the numbers from $n+1$ to $2n-1$ inclusive is chosen, then its difference from $3n$ will be greater than n by less than $2n$. Hence the $n+1$ chosen numbers other than $3n$ come from the n pairs $(1, 2n)$, $(2, 2n+1)$, $(3, 2n+2), \ldots, (n, 3n-1)$. By the Pigeonhole Principle, both members of some pair are among the chosen numbers, their difference strictly between n and $2n$.

Second Solution Mark $3n$ evenly distributed points on a circle and label them from 1 to $3n$ in clockwise order. If the distance along the circle of two points is greater than one-third of the circle but less than two-thirds, then the difference between their labels is greater than n but less than $2n$. Note that it makes no difference whether distance is measured clockwise or counter-clockwise. Two such points are said to determine a desirable partition. We now choose a subset of the points so that no two of them determine a desirable partition. We define a free arc as one whose endpoints are in the subset, but no points in its interior are in the subset. Let P be a chosen point and Q be the point diametrically opposite P. Then the arc of length one-third of the circle and with Q as its midpoint must be free. Hence there is a free arc of length at least one third of the circle. We cannot have a free arc whose length is greater than one-third of the circle and less than two-thirds, since its endpoints will determine a desirable partition. Suppose the longest free arc AB has length equal to one-third of the circle. Let C be the third point of trisection of the circle. Since AB is free, no points in its interior belong to the subset. Since A is chosen, no points in the interior of the arc BC belong to the subset. Since B is chosen, no points in the interior of the arc CA belong to the subset. Hence the subset can have at most 3 points. Since $n > 1$, we cannot get $n+2$ points. Finally, suppose the longest free arc has length at least two-thirds of the circle. Then there are $n+1$ points not in its interior, and we cannot get $n+2$ points.

Third Solution If we choose the number k, then we cannot choose any of $k+n+1, k+n+2, \ldots, k+2n-1$. To maximize the number of chosen ones, we may as well take 1. Then we cannot choose $n+2, n+3, \ldots, 2n$. Suppose we also take $n+1$. Then we cannot choose $2n+2, 2n+3, \ldots, 3n$. We can either take $2n+1$ for a total of 3 numbers, or all of $2, 3, \ldots, n$ for a total of $n+1$ numbers. Both fall short of the target of $n+2$ numbers. Suppose we do not take $n+1$. Let $k \geq 1$ be the largest chosen number under $n+1$. Then we cannot choose any of $k+1, k+2, \ldots, k+2n-1$. With $2n-1$ of the $3n$ numbers ruled out, we can choose at most $n+1$ numbers, again falling short.

3.6 Problem Set: Algebra

Problem 1961.2

Prove that if a, b and c are positive real numbers, each less than 1, then the products $(1-a)b$, $(1-b)c$ and $(1-c)a$ cannot all be greater than $\frac{1}{4}$.

First Solution For any real number x, $(x-\frac{1}{2})^2 \geq 0$, so that $\frac{1}{4} \geq x(1-x)$. If $0 < x < 1$, then $x(1-x) > 0$. Hence the product of $(1-a)b$, $(1-b)c$ and $(1-c)a$ is $a(1-a)b(1-b)c(1-c) \leq (\frac{1}{4})^3$, so that it is impossible for each factor to exceed $\frac{1}{4}$.

Second Solution It follows from the Arithmetic-Geometric Mean Inequality that $\sqrt{x(1-x)} \leq \frac{x+(1-x)}{2} = \frac{1}{2}$ for any real number x such that $0 < x < 1$. Hence $x(1-x) \leq \frac{1}{4}$. By cyclic symmetry, we may assume that $a \geq b$. Then $(1-a)b \leq (1-a)a \leq \frac{1}{4}$.

Problem 1957.3

Let n be a positive integer and $\langle a_1, a_2, \ldots, a_n \rangle$ be any permutation of $1, 2, \ldots, n$. Determine the maximum value of the expression $|a_1 - 1| + |a_2 - 2| + \cdots + |a_n - n|$.

First Solution We may replace $|a_k - k|$ by $a_k - k$ if $a_k \geq k$ and by $k - a_k$ otherwise. This yields a sum of $2n$ terms, in which each of the numbers from 1 to n appears twice. Exactly n of the terms are prefixed by minus signs. To maximize the sum, the smallest n terms should be prefixed by minus signs. For $n = 2m$, this can be achieved by taking $a_k = m + k$ for $1 \leq k \leq m$ and $a_k = k - m$ for $m+1 \leq k \leq 2m$. The sum is $S = 2m^2$. For $n = 2m+1$, this can be achieved by taking $a_k = m + k + 1$ for $1 \leq k \leq m$, $a_{m+1} = m+1$ and $a_k = k - m - 1$ for $m+2 \leq k \leq 2m+1$. The sum is $S = 2m(m+1)$. Combining the two cases, we have $S = \lfloor \frac{n^2}{2} \rfloor$.

Second Solution We shall prove that the optimal result is obtained if we take $a_k = n + 1 - k$ for $1 \leq k \leq n$. Let i be the smallest index for which $a_i \neq n + 1 - i$. Then $a_j = n + 1 - i$ for some $j > i$, so that $a_i < a_j$. We claim that switching a_i with a_j will not decrease the sum. In other words,

$$|i - a_i| + |j - a_j| \leq |i - a_j| + |j - a_i|.$$

Note that $|x-y| = x+y-2\min\{x,y\}$. Hence the desired result is

$$\min\{i,a_j\} + \min\{j,a_i\} \le \min\{i,a_i\} + \min\{j,a_j\}.$$

Suppose $i \le a_j$ and $a_i \le j$. Then the left side of the displayed equation is $i + a_i$ while the right side cannot be less than that. If $a_j < i$, then both sides are $a_i + a_j$. If $j < a_i$, then both sides are $i + j$. This justifies the claim. We now calculate the value of

$$S = |1-n| + |2-(n-1)| + \cdots + |(n-1)-2| + |n-1|.$$

For n odd, the terms decrease from $n-1$ by steps of 2 until 0, and then increase to $n-1$ by steps of 2. Hence

$$S = 2(2+4+\cdots+(n-1)) = \frac{n-1}{2}(2+(n-1)) = \frac{n^2-1}{2}.$$

For n even, the terms decrease from $n-1$ by steps of 2 until 1, and then increase from a second copy of 1 to $n-1$ by steps of 2. Hence

$$S = 2(1+3+\cdots+(n-1)) = \frac{n}{2}(1+(n-1)) = \frac{n^2}{2}.$$

Problem 1950.3

Let a_1, b_1, c_1, a_2, b_2 and c_2 be real numbers such that for any integers x and y, at least one of $a_1x + b_1y + c_1$ and $a_2x + b_2y + c_2$ is an even integer. Prove that either all of a_1, b_1 and c_1 are integers or all of a_2, b_2 and c_2 are integers.

First Solution Putting in $(x,y) = (0,0)$, $(1,0)$ and $(-1,0)$, one of c_1 and c_2, one of $a_1 + c_1$ and $a_2 + c_2$ and one of $-a_1 + c_1$ and $-a_2 + c_2$ are even integers. By the Pigeonhole Principle, we may assume that two of these even integers come from numbers with subscripts 1. If they are c_1 and $a_1 + c_1$, then $a_1 = (a_1 + c_1) - c_1$ is also an even integer. If they are c_1 and $-a_1 + c_1$, then $a_1 = c_1 - (-a_1 + c_1)$ is also an even integer. If they are $a_1 + c_1$ and $-a_1 + c_1$, then both $a_1 = \frac{(a_1+c_1)-(-a_1+c_1)}{2}$ and $c_1 = \frac{(a_1+c_1)+(-a_1+c_1)}{2}$ are integers. Similarly, we can show that either both b_1 and c_1 are integers, or both b_2 and c_2 are integers. In the former case, there is nothing further to prove. In the latter case, put $(x,y) = (1,1)$. Then the desired conclusion follows since either $a_1 + b_1 + c_1$ or $a_2 + b_2 + c_2$ is an even integer.

Second Solution Consider any nine lattice points in a 3×3 configuration. Color (x, y) red if $a_1x + b_1y + c_1$ is an even integer, and blue otherwise. By the Pigeonhole Principle, at least five of these nine lattice points have the same color, say red. Then two of them are in the same row and two are in the same column. Suppose the two in the same row are (x_1, y) and (x_2, y), where the difference $d = x_2 - x_1$ is either 1 or 2. Then $da_1 = (a_1x_2 + b_1y + c_1) - (a_1x_1 + b_1y + c_1)$ is an even integer. Hence a_1 is an integer. Suppose the two red lattice points in the same column are (x, y_1) and (x, y_2), where the difference $d = y_2 - y_1$ is either 1 or 2. Then $db_1 = (a_1x + b_1y_2 + c_1) - (a_1x + b_1y_1 + c_1)$ is an even integer. Hence b_1 is an integer. It follows that c_1 is also an integer.

Problem 1959.1

Prove that

$$\frac{x^n}{(x-y)(x-z)} + \frac{y^n}{(y-z)(y-x)} + \frac{z^n}{(z-x)(z-y)}$$

is an integer, where x, y and z are distinct integers and n is a non-negative integer.

First Solution The given expression is equal to

$$\frac{x^n(z-y) + y^n(x-z) + z^n(y-x)}{(x-y)(y-z)(z-x)}$$

and we need to prove that the numerator is divisible by the common denominator. The former is equal to 0 when $n = 0$ or 1. For $n \geq 2$, we have

$$\begin{aligned}
&x^n(z-y) + y^n(x-z) + z^n(y-x) \\
&\quad = x^n(z-y) - x(z^n - y^n) + yz(z^{n-1} - y^{n-1}) \\
&\quad = (z-y)(x^n - x(z^{n-1} + z^{n-2}y + \cdots + y^{n-1}) \\
&\qquad + yz(z^{n-2} + z^{n-3}y + \cdots + y^{n-2})).
\end{aligned}$$

Consider the second factor as a polynomial f in x, with y and z treated as unequal integral constants. Setting $x = y$, the displayed expression vanishes. Since $z \neq y$, we have $f = 0$. It follows that $f = (x-y)g$ for some polynomial g in x. Setting $x = z$, the displayed expression vanishes. Since $z \neq y$, $x \neq y$ so that $g = 0$. It follows that the displayed expression is divisible by $(z-y)(x-z)(y-x)$.

Second Solution The given expression is equal to

$$\frac{x^n(z-y)+y^n(x-z)+z^n(y-x)}{(x-y)(y-z)(z-x)}$$

and we need to prove that the numerator is divisible by the common denominator. The former is equal to 0 when $n=0$ or 1. For $n\geq 2$, we have

$$\begin{aligned}
&x^n(z-y)+y^n(x-z)+z^n(y-x)\\
&\quad = x^n((x-y)+(z-x))-y^n(z-x)-z^n(x-y)\\
&\quad = (z-x)(x^n-y^n)-(x-y)(z^n-x^n)\\
&\quad = (z-x)(x-y)(y-z)(x^{n-2}+x^{n-3}(y+z)+\cdots\\
&\qquad +(y^{n-2}+y^{n-3}z+\cdots+z^{n-2})).
\end{aligned}$$

Third Solution Let

$$P_n(x,y,z)=\frac{x^n(z-y)+y^n(x-z)+z^n(y-x)}{(x-y)(y-z)(z-x)}.$$

We shall prove by induction that $P_n(x,y,z)$ are polynomials in x, y and z with integral coefficients for all $n\geq 0$. This is certainly true for $n=0$ and 1 as $P_0(x,y,z)=P_1(x,y,z)=0$. Suppose it is true for some $n\geq 1$. Then

$$\begin{aligned}
P_{n+1}(x,y,z)-zP_n(x,y,z) &= \frac{x^{n+1}-zx^n}{(x-y)(x-z)}+\frac{y^{n+1}-zy^n}{(y-z)(y-x)}\\
&= \frac{x^n-y^n}{x-y}.
\end{aligned}$$

It follows that

$$P_{n+1}(x,y,z)=zP_n(x,y,z)+(x^{n-1}+x^{n-2}y+\cdots+y^{n-1}).$$

The desired result follows from the induction hypothesis.

Fourth Solution Let

$$P_n(x,y,z)=\frac{x^n(z-y)+y^n(x-z)+z^n(y-x)}{(x-y)(y-z)(z-x)}.$$

Both $P_0(x,y,z)$ and $P_1(x,y,z)$ vanish while $P_2(x,y,z)=1$. Note that

$t^n(t-x)(t-y)(t-z)=t^{n+3}-(x+y+z)t^{n+2}+(yz+zx+xy)t^{n+1}-xyzt^n$.

Hence $t^{n+3}=(x+y+z)t^{n+2}-(yz+zx+xy)t^{n+1}+xyzt^n$ whenever $t=x$, y or z. Now

$$\begin{aligned}
P_{n+3}(x,y,z) &= (x+y+z)P_{n+2}(x,y,z)\\
&\qquad -(yz+zx+xy)P_{n+1}(x,y,z)+xyzP_n(x,y,z).
\end{aligned}$$

It follows that $P_n(x,y,z)$ are polynomials in x, y and z with integral coefficients for all $n\geq 0$.

3.7 Problem Set: Geometry

Problem 1958.3

$ABCDEF$ is a convex hexagon in which opposite edges are parallel. Prove that triangles ACE and BDF have equal area.

First Solution Let BE and CF meet each other at P and meet AD at R and Q respectively. These three points may coincide. Divide triangle ACE into four triangles, PCE, QCA, RAE and PQR. Divide triangle BDF into four triangles, PBF, QFD, RDB and PQR.

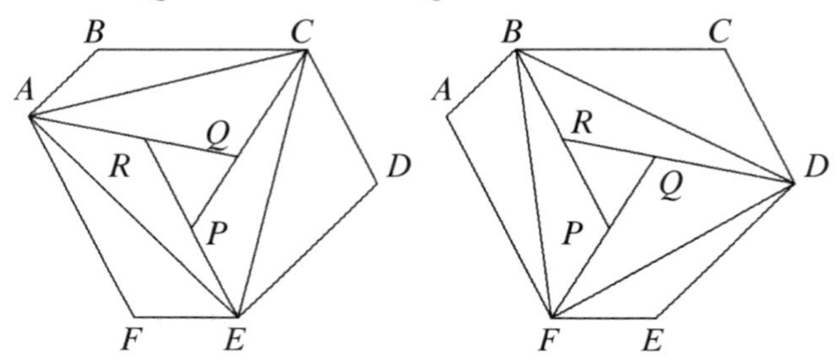

Since AF is parallel to CD, triangles CAF and DAF have the same area. Taking away the common triangle QAF, triangles QCA and QFD also have the same area. Similarly, we can prove that triangles PCE and PBF have the same area, as have triangles RAE and RDB. It follows that triangles ACE and BDF have the same area.

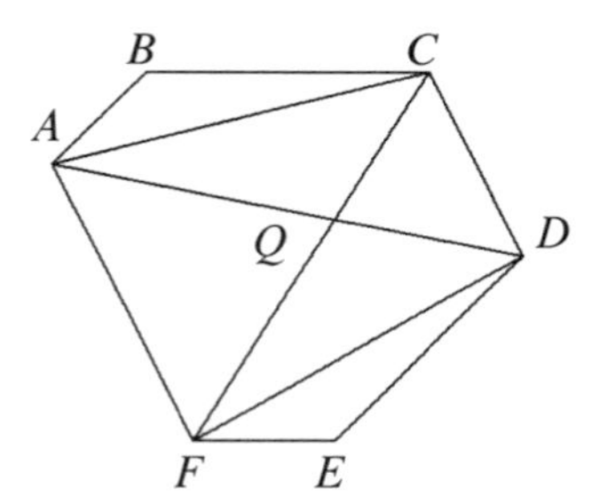

Second Solution Complete the parallelograms $ABCU$, $CDEV$ and $EFAW$. Since the opposite sides of $ABCDEF$ are parallel, A, C and E lie on the extensions of the sides of triangle UVW. Divide triangle ACE four triangles, ACU, CEV, EAW and UVW. The first three are equal in area respectively to triangles ABC, CDE and EFA. It follows that the area of triangle ACE is the average of the areas of $ABCDEF$ and UVW. Similarly, the area of BDF is the average of the areas of $ABCDEF$ and XYZ, where $BCDX$, $DEFY$ and $FABZ$ are parallelograms. Now $UV = DE - AB = YZ$.

Similarly, $VW = ZX$ and $WU = XY$. Hence WUV and XYZ are congruent and have the same area. It follows that triangles ACE and BCF also have the same area.

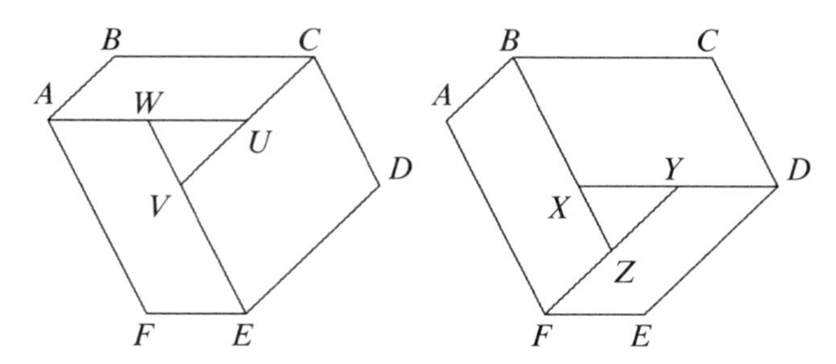

Third Solution For any triangle XYZ, denote by $s(XYZ)$ it signed area, that is, $s(XYZ)$ is the area of XYZ if X, Y and Z are in counterclockwise order, and the negative of the area otherwise. We claim that

$$s(XYZ) = s(OXY) + s(OYZ) + s(OZX)$$

for any point O on the plane of XYZ. If O is inside XYZ, this follows from the consideration of ordinary area. If O is in the infinite region bounded by one side of XYZ, say ZX, and the extensions of the other two sides, then the third term on the right is negative, but the result still holds.

If O is in the infinite region bounded by the extensions of two sides of XYZ, say ZX and YX, then the first two terms on the right are negative, but again the result holds. This justifies the claim.

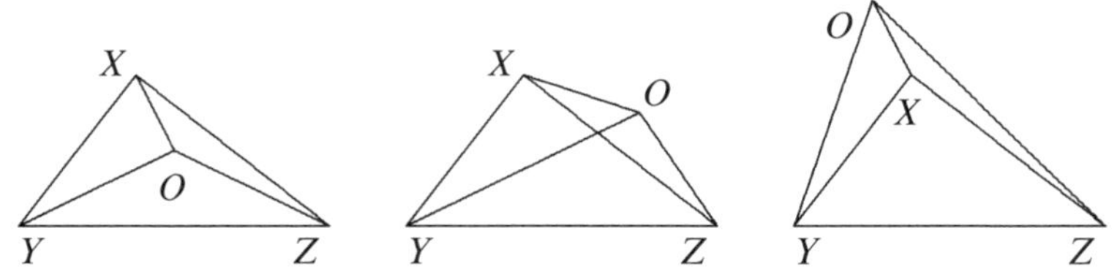

We now return to the problem at hand. We have AB parallel to DE. Since ABD and ABE are in the same orientation, $s(ABE) = s(ABD)$. For any point O in the plane of $ABCDEF$, we have

$$s(OAB) + s(OBE) + s(OEA) = s(OAB) + s(OBD) + s(ODA).$$

In a similar way, from the fact that $s(CDA) = s(CDF)$ and $s(EFC) = s(EFB)$, we have

$$s(OCD) + s(ODA) + s(OAC) = s(OCD) + s(ODF) + s(OFC)$$

and

$$s(OEF) + s(OFC) + s(OCE) = s(OEF) + s(OFB) + s(OBE).$$

Adding all three equalities yields

$$s(OEA) + s(OAC) + s(OCE) = s(OBD) + s(ODF) + s(OFB)$$

or $s(ACE) = s(BDF)$. The desired result follows immediately.

Problem 1951.1

$ABCD$ is a square of side a. E is a point on BC with $BE = \frac{a}{3}$. F is a point on DC extended with $CF = \frac{a}{2}$. Prove that the point of intersection of AE and BF lies on the circumcircle of $ABCD$.

First Solution Let M be the point of intersection of AE and BF, and let the extensions of AM and DC meet at G. Let UV be such that C is the midpoint of AV and of BU. Then triangles AVG and BUV are similar since both are right triangles whose legs are in the ratio 2:1. Hence $\angle MAC = \angle MBC$, so that M lies on the circumcircle of $ABCD$.

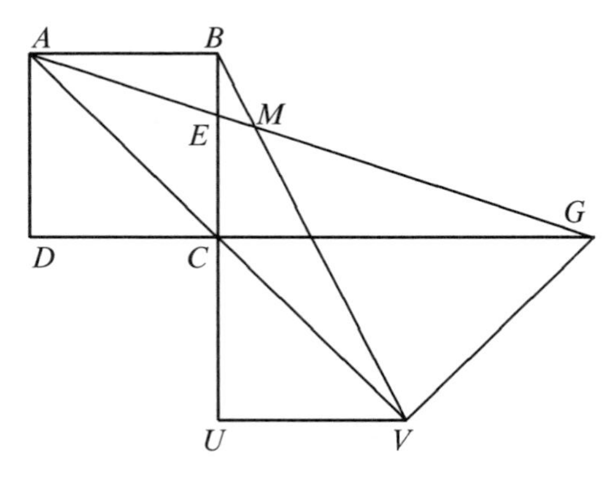

Second Solution Let M be the point of intersection of AE and BF, and let the extensions of AM and DC meet at G. Let UV be such that C is the midpoint of AV and of BU. Now AUG is an isosceles right triangle. Hence $\angle UAG = 45°$. Since $ABVU$ is a parallelogram,

$$\angle BMA = \angle MAU = 45° = \angle BDA.$$

Hence M lies on the circumcircle of $ABCD$.

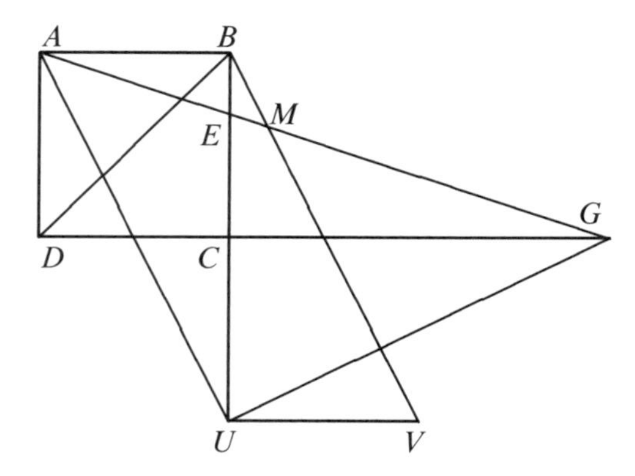

Third Solution Let M be the point of intersection of AE and BF. Let G be the point of intersection of AM and DC, and let H be the intersection of CM and AB. Since triangles ABE and GDA are similar, $GD = 3a$. Hence $\frac{BH}{BA} = \frac{FC}{FG} = \frac{1}{3}$. Hence $BH = \frac{a}{3} + BE$. Rotating triangle ABE counterclockwise 90° about B, it coincides with triangle CBH. Hence AE and CH are perpendicular, so that $\angle AMC = 90^\circ$. It follows that M lies on the circumcircle of $ABCD$ which has AC as a diameter.

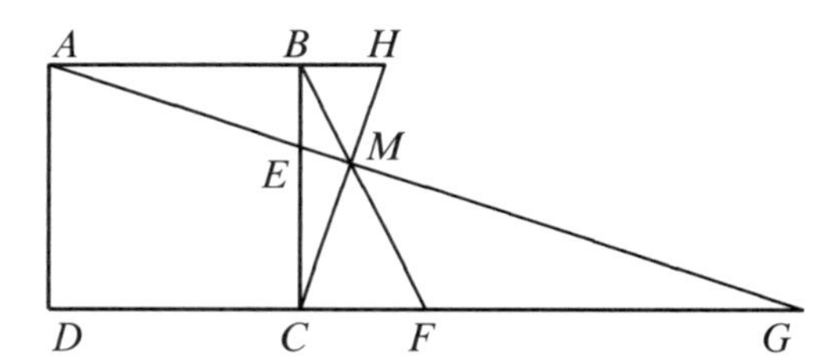

Problem 1949.2

Let P be any point on the base of a given isosceles triangle. Let Q and R be the intersections of the equal sides with lines drawn through P parallel to these sides. Prove that the reflection of P about the line QR lies on the circumcircle of the given triangle.

First Solution Let ABC be a triangle with $AB = AC$, so that we have $\angle ABC = \angle ACB$. Let P, Q and R be on sides BC, CA and AB respectively. Let D be the point of reflection of P across QR. Then $RP = RD$ and $QP = QD$. Moreover, $AQPR$ is a parallelogram, so that $AR = QP = QD$ and $AQ = PR = DR$. It follows that triangles QAD and RDA are congruent, so that $\angle QAD = \angle RDA$. Since RBP is similar to ABC,

$RB = RP = RD$, so that we have $\angle RBD = \angle RDB$. Hence

$$\begin{aligned}\angle ADB + \angle BCA &= \angle RDA + \angle RDB + \angle ACB \\ &= \angle QAD + \angle RBD + \angle ABC \\ &= \angle CAD + \angle DBC.\end{aligned}$$

Since the sum of all four angles is 360°, the sum of each pair is 180°, so that $ADBC$ is a cyclic quadrilateral.

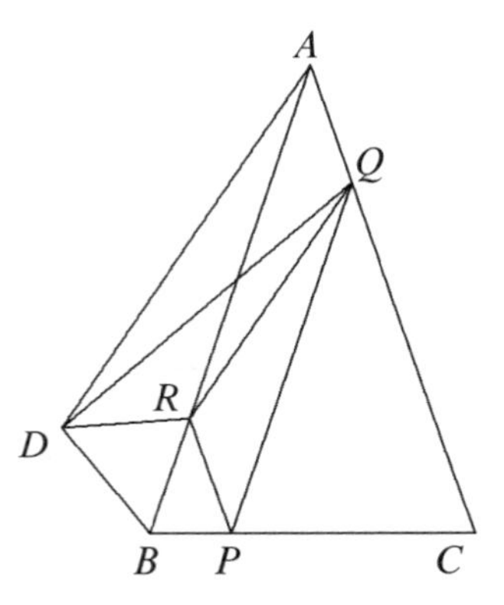

Second Solution Let ABC be a triangle with $AB = AC$. Let P, Q and R be on sides BC, CA and AB respectively. Let D be the point of reflection of P across QR. Then we have

$$\angle CAB = \angle RPQ = \angle RDQ.$$

Hence $AQRD$ is a cyclic quadrilateral. It follows that $\angle AQD = \angle ARD$ and $\angle DQC = \angle DRB$. Note that we have $DQ = PQ = CQ$ and $DR = PR = BR$. Hence the isosceles triangles DQC and DRB are similar, so that $\angle BDR = \angle CDQ$. Now

$$\begin{aligned}\angle BDC &= \angle BDQ - \angle CDQ \\ &= \angle BDQ - \angle BDR \\ &= \angle RDQ \\ &= \angle BAC.\end{aligned}$$

Since A and D are on the same side of BC, $ADBC$ is a cyclic quadrilateral.

Third Solution Let ABC be a triangle with $AB = AC$. Let P, Q and R be on sides BC, CA and AB respectively. Let D be the point of reflection of P across QR. Then $RP = RD$ and $QP = QD$. Moreover, both RBP

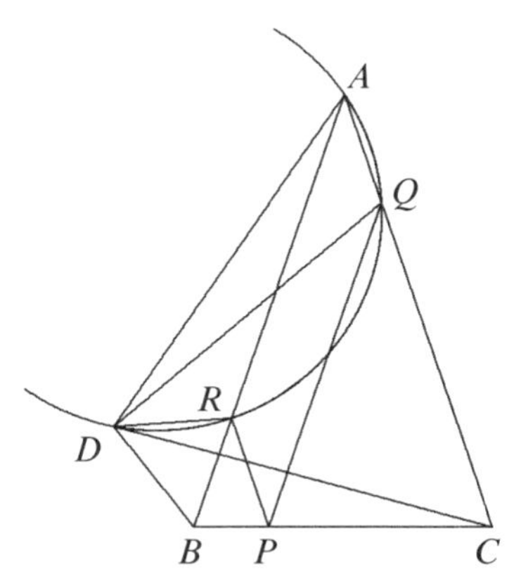

and QPC are similar to ABC. Hence $RB = RP = RD$ and $QC = QP = QD$. It follows that R is the circumcenter of DBP and Q is the circumcenter of DPC. We now have $2\angle CDP = \angle CQP$ and $2\angle PDB = \angle BRP$. Since $\angle CQP = \angle CAB = \angle BRP$, $\angle CDB = \angle CDP + \angle PDB = \angle CAB$. Since A and D are on the same side of BC, $ADBC$ is a cyclic quadrilateral.

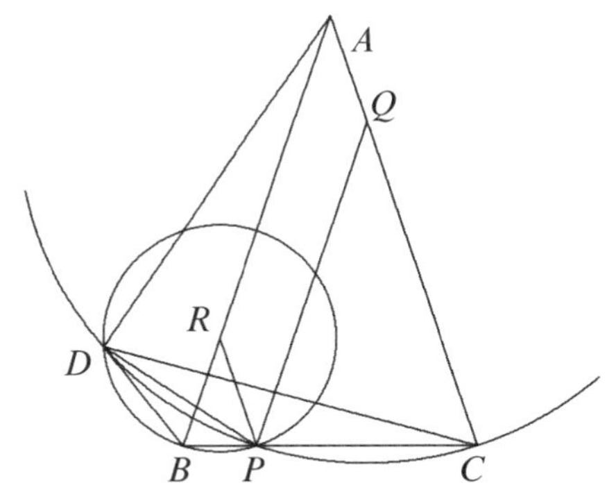

Fourth Solution Let ABC be a triangle with $AB = AC$. Let P, Q and R be on sides BC, CA and AB respectively. Let D be the point of reflection of P across QR. Then $RP = RD$ and $QP = QD$. Moreover, $AQPR$ is a parallelogram, so that $AR = QP = QD$ and $AQ = PR = DR$. It follows that triangles QAD and RDA are congruent, so that $ADRQ$ is an isosceles trapezoid. Let O be the circumcenter of ABC. Rotate the segment AB about O until the image of A coincides with C and the image of B coincides with A. Then the image of R coincides with Q. It follows that $OQ = OR$, so that O lies on the perpendicular bisector of QR. Hence O also lies on the perpendicular bisector of AD, so that D indeed lies on the circumcircle of ABC.

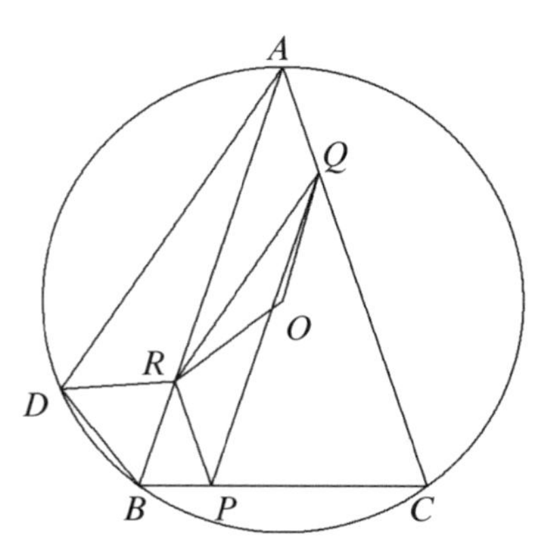

Fifth Solution We prove a more general result, where ABC is an arbitrary triangle. P is any point on BC. Q and R are points on CA and AB such that $QP = QC$ and $RP = RB$, respectively. Let the line through A parallel to BC cut the extensions of PQ and PR at Y and X, respectively. Then we have $\angle CAY = \angle ACP = \angle YPC$ and $\angle BAX = \angle ABP = \angle XPB$. It follows that $\angle CAB = \angle YPX$. Now $AC = PY$ and $AB = PX$. Hence triangles ABC and PXY are congruent, so that they have equal circumradii. The power of the point Q with respect to the two circles are $QA \cdot QC = QY \cdot QP$. Hence it lies on the radical axis of the two circles. Similarly, so does R. Since the two circles are congruent, they are symmetric about QR, and the point D which is the image of P across QR lies on the circumcircle of ABC.

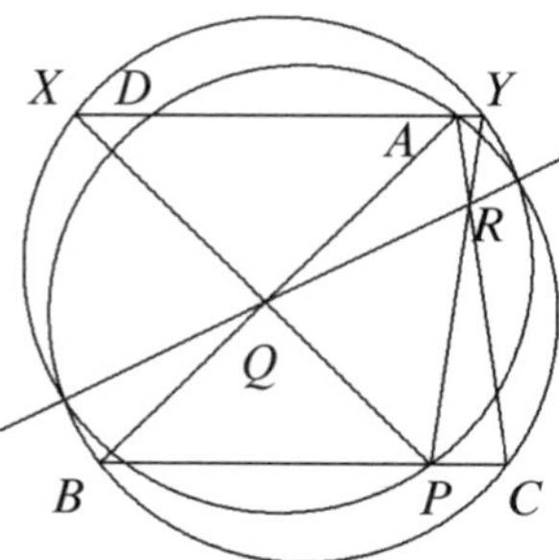

Problem 1955.3

The three vertices of a certain triangle are lattice points. There are no other lattice points on its perimeter but there is exactly one lattice point in its interior. Prove that this lattice point is the centroid of the triangle.

First Solution Let ABC be the lattice triangle and P be the sole lattice point inside. Complete the parallelograms $ABDC$, $BCEA$ and $CAFB$. Let

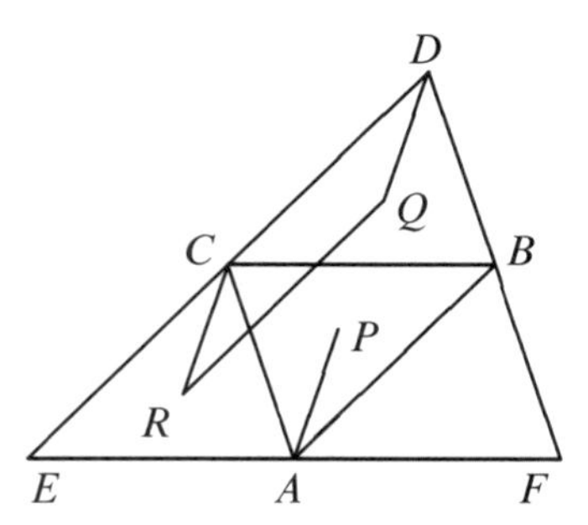

Q, R and S be the respective images of reflection of P across the midpoints of BC, CA and AB. Then they are sole lattice points inside triangles DBC, ECA and FAB respectively. Consider the image of reflection of D across Q. It is a lattice point inside DEF, and since P, Q, R and S are the only lattice points there, it must coincide with one of them. Clearly it cannot be Q. Suppose it is R. Now CR is the reflection of AP across the midpoint of CA, and AP is in turn the reflection of DQ across the midpoint of BC. Hence CR and DQ are equal and parallel, so that $CRQD$ is a parallelogram. Hence it is impossible for R to be the image of reflection of D across Q. In an analogous way, we can prove that it is not S either. Hence D, P and Q are collinear. Since PQ passes through the midpoint of BC, DP passes through the midpoint A of EF, and AP is part of a median of triangle ABC. Similarly, so are BP and CP, so that P is the centroid of ABC.

Second Solution Let ABC be the lattice triangle and P be the sole lattice point inside. Complete the parallelogram $BCEA$ and let R be the image of reflection of P across the midpoint of CA. Then it is the sole lattice point inside triangle ECA. Let L, M and N be the midpoints of CE, EA and AB respectively. Then R is inside AMN, CNL or $LEMN$. Suppose it is inside AMN. Consider the image of reflection of A across R. It is another lattice point inside ECA, which is impossible. Similarly, R is not inside CNL.

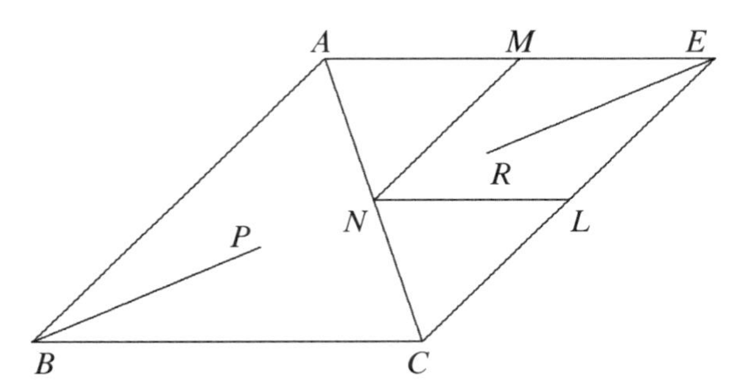

Hence it is inside $LEMN$. Consider the image of reflection of E across R. It is another lattice point inside $BCEA$. Hence it must be P. Similarly, the image of reflection of B across P is R. Hence P and R lies on the diagonal BE of $BCEA$, and we have $BP = PR = RE$. Now BE also passes through M, and since $BM = EM$, we have $PM = RM$. It follows that $AP = 2PM$, and P is indeed the centroid of ABC.

Third Solution Let ABC be the lattice triangle and P be the sole lattice point inside. By Pick's Formula, the area of each of triangles PBC, PCA and PAB is $\frac{1}{2}$. Hence the extensions of AP, BP and CP will bisect the opposite sides, so that P is indeed the centroid of ABC.

Fourth Solution Let ABC be the lattice triangle and P be the sole lattice point inside. The medians AK, BN and CH are concurrent at the centroid G, and divides ABC into six triangles. P must be inside one of them, and by symmetry, we may assume it is GBK. Let T be the image of reflection of B across G. Then the image X of reflection of B across P is a lattice point inside triangle TBC. It must be inside triangle TNC since there are no other lattice points inside ABC. Now the image Y of reflection of X across N is a lattice point inside NAG. Since P is the only lattice point inside ABC, Y must coincide with P. However, the only common point of GBK and NAG is G. Hence P coincides with the centroid G.

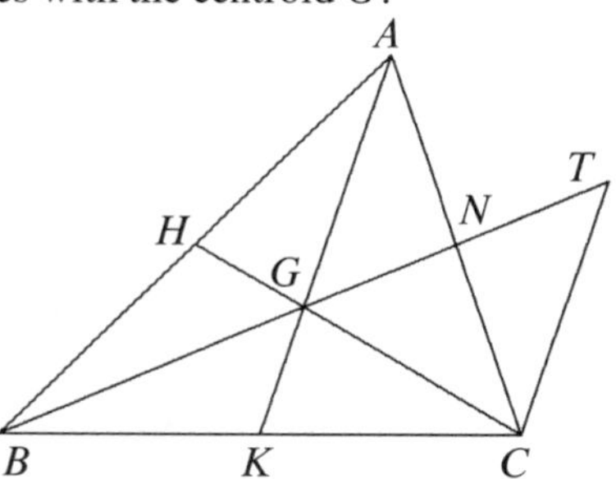

3.8 Problem Set: Tangent Lines and Circles

Problem 1950.2

Three circles k_1, k_2 and k_3 on a plane are mutually tangent at three distinct points. The point of tangency of k_1 and k_2 is joined to the other two points of tangency. Prove that these two segments or their extensions intersect k_3 at the endpoints of one of its diameters.

Solution Let the points be labeled as shown in the diagram below. Note that

$$\angle O_1T_3T_2 = \angle O_1T_2T_3 = \angle O_3T_2B = \angle O_3BT_2.$$

Hence O_3B is parallel to T_3O_1. Similarly, O_3A is parallel to T_3O_2. Since T_3, O_1 and O_2 are collinear, so are A, B and O_3. Since A and B are distinct points, AB is a diameter of k_3.

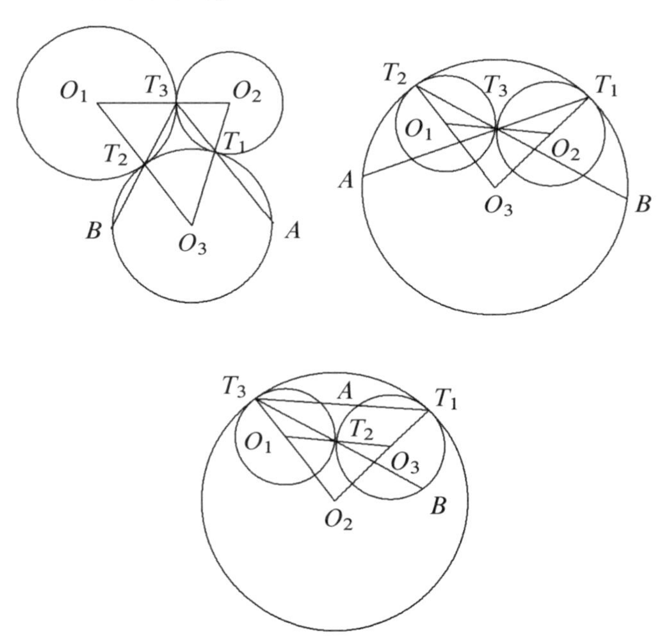

Problem 1952.1

The centers of three mutually disjoint circles lie on a line. Prove that if a fourth circle is tangent to each of them, then its radius is at least the minimum of the radii of the other three circles.

First Solution If the fourth circle is inside one of the other three, then it cannot be tangent to the remaining two. If one of the other three circles is inside the fourth one, then the radius of the fourth circle is not smaller than the radius of the circle inside it. Henceforth we may restrict our attention to external tangencies only. Let the collinear centers be O_1, O_2 and O_3 in that order, and let the center of the fourth circle be O. Let the respective radii be

r_1, r_2, r_3 and r. By the Triangle Inequality,

$$\begin{aligned} &(r + r_1) + (r + r_3) \\ &= OO_1 + OO_3 \\ &\geq O_1O_3 \\ &\geq r_1 + 2r_2 + r_3. \end{aligned}$$

Hence $r \geq r_2$.

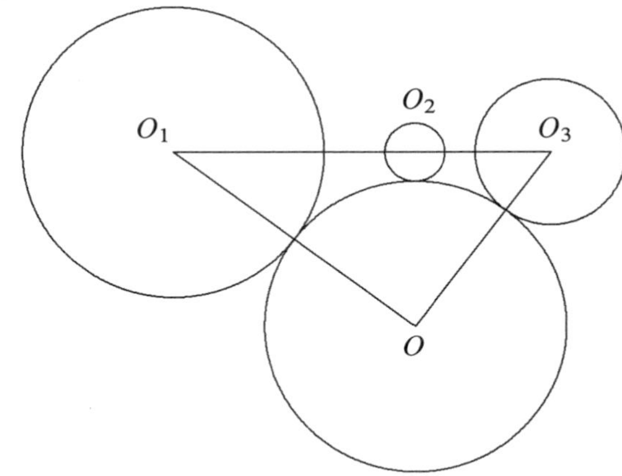

Second Solution If the fourth circle is inside one of the other three, then it cannot be tangent to the remaining two. If one of the other three circles is inside the fourth one, then the radius of the fourth circle is not smaller than the radius of the circle inside it. Henceforth we may restrict our attention to external tangencies only. Let the collinear centers be O_1, O_2 and O_3 in that order, and let the center of the fourth circle be O. Let the respective radii be r_1, r_2, r_3 and r. Let T_1 and T_3 be the points of tangency of the fourth circle with the first and the third, and let A_1 and A_3 be the points of intersection of the segment O_1O_3 with the first and third circle, respectively. Then A_1A_3 is the shortest distance from any point on the first circle to any point on the third circle. Hence

$$\begin{aligned} 2r_2 &\leq A_1A_3 \\ &\leq T_1T_3 \\ &\leq OT_1 + OT_3 \\ &= 2r, \end{aligned}$$

so that $r \geq r_2$.

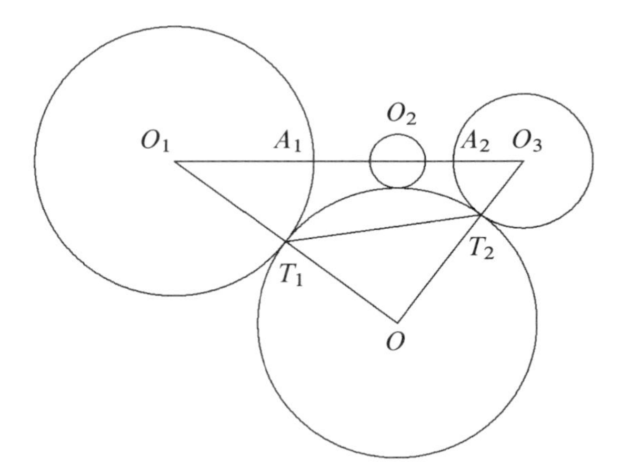

Third Solution If the fourth circle is inside one of the other three, then it cannot be tangent to the remaining two. If one of the other three circles is inside the fourth one, then the radius of the fourth circle is not smaller than the radius of the circle inside it. Henceforth we may restrict our attention to external tangencies only. Let the collinear centers be O_1, O_2 and O_3 in that order, and let the center of the fourth circle be O. Let the respective radii be r_1, r_2, r_3 and r. Note that $\angle OO_2O_1 + \angle OO_2O_3 = 180°$. By symmetry, we may assume that $\angle OO_2O3 \geq 90° \geq \angle O_2OO_3$. Hence

$$\begin{aligned} r + r_3 &= OO_3 \\ &\geq O_2O_3 \\ &\geq r_2 + r_3, \end{aligned}$$

so that $r \geq r_2$.

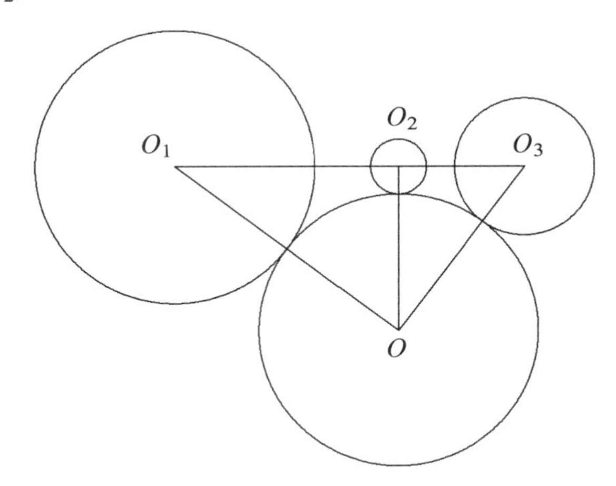

Fourth Solution If the fourth circle is inside one of the other three, then it cannot be tangent to the remaining two. If one of the other three circles is inside the fourth one, then the radius of the fourth circle is not smaller than the radius of the circle inside it. Henceforth we may restrict our attention to external tangencies only. Let the collinear centers be O_1, O_2 and O_3 in that order, and let the center of the fourth circle be O. Let the respective radii be r_1, r_2, r_3 and r. Suppose to the contrary that $r < r_2$. Increase the radii of the first and third circles by r_2. Then O must lie inside both expanded circles. However, considerations along the line O_1O_3 makes it clear that the two expanded circles cannot have any common interior points. This contradiction shows that $r \geq r_2$.

Fifth Solution As in the Fourth Solution, we may restrict our attention to external tangencies only. Let the collinear centers be O_1, O_2 and O_3 in that order, and let the center of the fourth circle be O. Let the respective radii be r_1, r_2, r_3 and r. Draw the tangents to the second circle perpendicular to O_1O_3. This infinite strip separates the first and third circles. Since the fourth circle is tangent to both, its diameter cannot be less than the width of the strip. It follows that $r \geq r_2$.

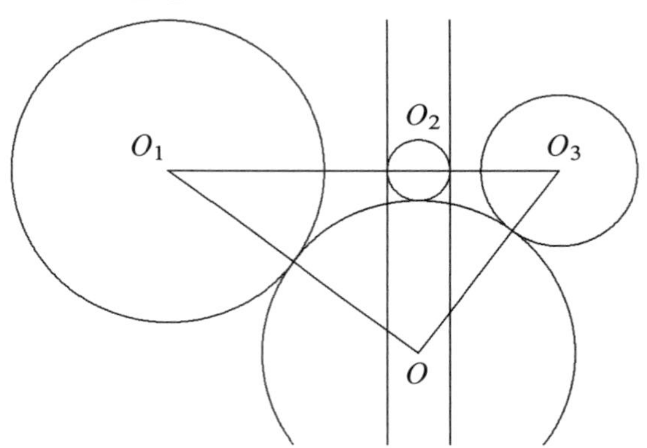

Problem 1960.3

In the square $ABCD$, E is the midpoint of AB, F is on the side BC and G on CD such that AG and EF are parallel. Prove that FG is tangent to the incircle of the square.

First Solution More generally, let $ABCD$ be a rhombus and let E be the point of tangency of the incircle of $ABCD$ with AB. If $ABCD$ is a square, E will be the midpoint of AB. Furthermore, we allow F to be on the extension of BC or G to be on the extension of DC. Let F be chosen and let the second tangent from F to the incircle of $ABCD$ cut the line DC at P. We shall prove that AP is parallel to EF, so that G and P coincide. Consider the excircle of triangle CFP opposite C, tangent to CD at Q. Let L be the point of tangency of CD with the incircle of $ABCD$. We have $PQ = CL = AE$, so that $APQE$ is a parallelogram, so that AP is parallel to EQ. Finally, F is the center of homothety of the two circles, so that E, F and Q are collinear. Hence AP is parallel to EF, and P coincides with G.

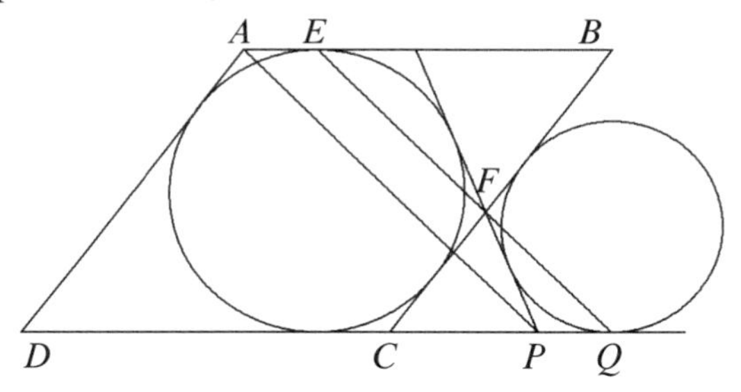

Second Solution Let I be the center of $ABCD$. Since ADG and FBE are similar triangles, $\frac{DA}{DG} = \frac{BF}{BE}$. Since IBE and ADI are also similar to each other, $\frac{BE}{BI} = \frac{DI}{DA}$. It follows that we have $BF \cdot DG = BE \cdot DA = BI \cdot DI$ so that $\frac{BF}{BI} = \frac{DI}{DG}$. Since $\angle FBI = 45° = \angle IDG$, FBO and ODG are similar. Hence $\angle FIB = \angle IGD$. It follows that

$$\begin{aligned}\angle FIG &= 180° - \angle GID - \angle FIB \\ &= 180° - \angle GID - \angle IGD \\ &= \angle IDG \\ &= 45°.\end{aligned}$$

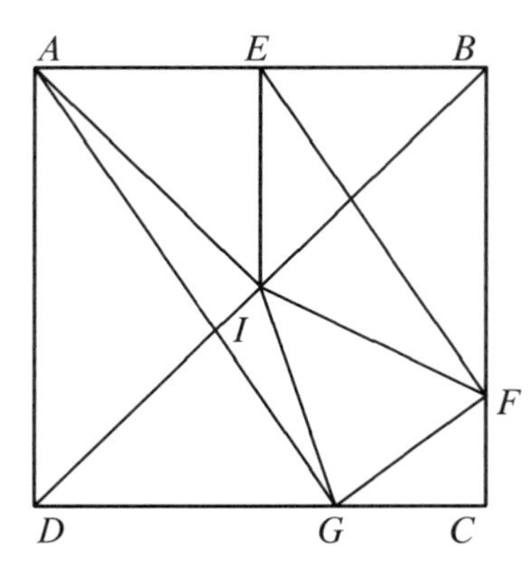

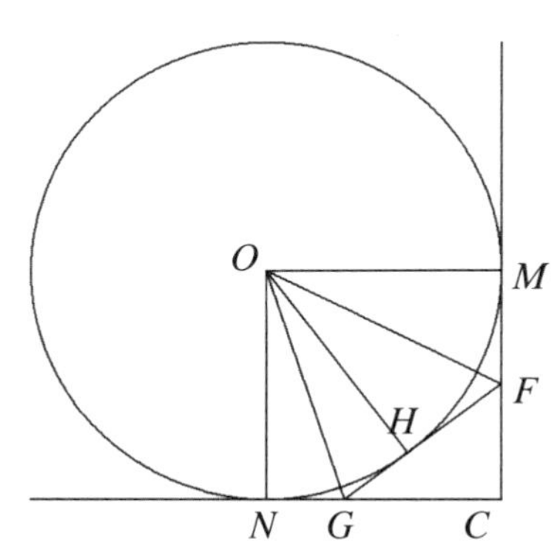

Let O be the excenter of triangle CFG opposite C and let H, M and N be the points of tangency of the excircle with the three sides. Then we have $\angle FOH = \angle FOM$ and $\angle GOH = \angle GON$, so that $\angle FOG = \frac{1}{2}\angle MON = 45°$. Now O lies on the bisector of $\angle FCG$. If it moves along this line towards C, $\angle FOG$ will increase. If it moves away from C instead, $\angle FOG$ will decrease. Hence O coincides with I, and FG is tangent to the incircle of $ABCD$.

Third Solution Let K and L be the respective midpoints of BC and CD. Then F lies on CK as otherwise G will lie on the extension of DC. Similarly, G lies on CL as otherwise F lies on the extension of BC. Take the side of $ABCD$ to be 2, and let $DG = x$. Since ADG and FBE are similar triangles, we have $\frac{BF}{BE} = \frac{DA}{DG}$ so that $FB = \frac{2}{x}$. Now

$$(FK + GL)^2 = \left(\frac{2}{x} - 1 + x - 1\right)^2 = x^2 - 4x + 8 - \frac{8}{x} + \frac{4}{x^2}.$$

Also,

$$FG^2 = CF^2 + CG^2 = \left(2 - \frac{2}{x}\right)^2 + (2 - x)^2 = x^2 - 4x + 8 - \frac{8}{x} + \frac{4}{x^2}.$$

It follows that $FK + GL = FG$. Let O be the excenter of triangle CFG opposite C and let H, M and N be the points of tangency of the excircle with the three sides. Then $FG = FH + HG = FM + GN$. Now O lies on the bisector of $\angle FCG$. If it moves along this line towards C, both FM and GN will decrease while FG remains constant. If it moves away from C instead, both FM and GN will increase. Hence M coincides with K and N with L, and O is the incenter of $ABCD$.

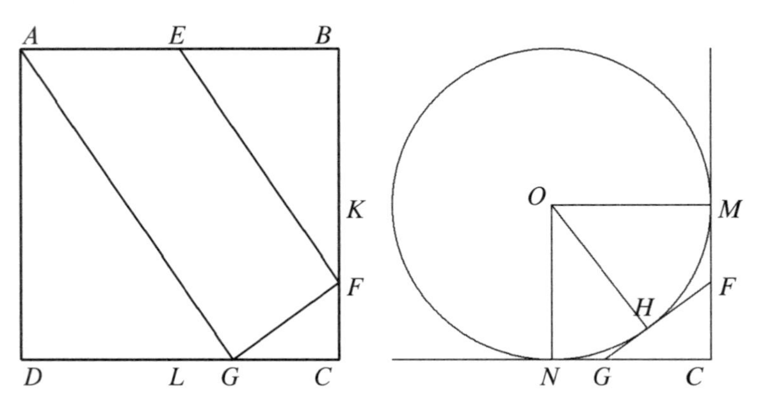

Problem 1961.3

Given two circles, exterior to each other, a common inner and a common outer tangents are drawn. The resulting points of tangency define a chord in each circle. Prove that the point of intersection of these two chords or their extensions is collinear with the centers of the circles.

First Solution Let the centers of the circles be O_1 and O_2. Let the external common tangent be A_1A_2 and the internal common tangent be B_1B_2, intersecting A_1A_2 at C. Let O_1O_2 intersect A_2B_2 at P_2 and the extension of A_1B_1 at P_1 (not shown in the diagram). Let CO_1 intersect A_1B_1 at D_1 and CO_2 intersect A_2B_2 at D_2. Now the kites $CA_1O_1B_1$ and $O_2A_2CB_2$ are similar, because each has a pair of opposite right angles, and $\angle A_2CB_2 = 180° - \angle B_1CA_1 = \angle A_1O_1B_1$. It follows that $\frac{CD_1}{CO_1} = \frac{D_2O_2}{CO_2}$. Now CO_1 and CO_2 are perpendicular since they are bisectors of supplementary angles. Hence CO_1 is parallel to A_2B_2 so that $\frac{D_2O_2}{CO_2} = \frac{P_2O_2}{O_1O_2}$. Similarly, CO_2 is parallel to A_1B_1 and $\frac{P_1O_2}{O_1O_2} = \frac{CD_1}{CO_1}$. It follows from $\frac{P_1O_2}{O_1O_2} = \frac{P_2O_2}{O_1O_2}$ that $P_1 = P_2$, so that O_1O_2, A_1B_1 and A_2B_2 are concurrent.

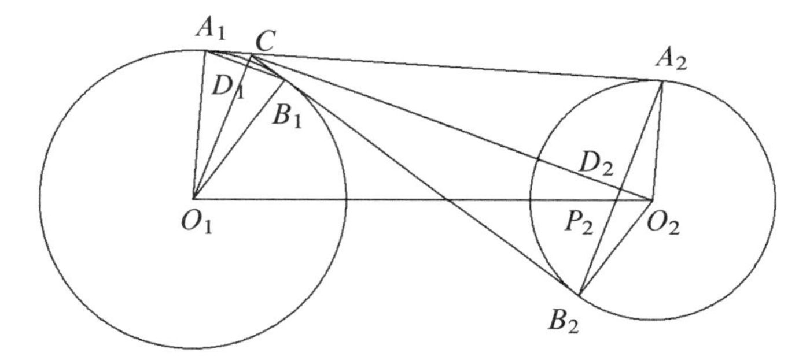

Second Solution Let the centers of the circles be O_1 and O_2. Let the external common tangent be A_1A_2 and the internal common tangent be B_1B_2, intersecting A_1A_2 at C. Draw the other internal common tangent Q_1Q_2, cutting A_1A_2 at N. Let M be the foot of perpendicular from N to O_1O_2. Draw the circle with diameter NO_1. It passes through A_1, Q_1 and M. Now

$$\angle O_1MB_1 = \angle O_1MQ_1 = \angle O_1NQ_1 = \angle O_1NA_1 = \angle O_1MA_1.$$

Hence A_1, B_1 and M are collinear. Draw the circle with diameter NO_2, and we can argue as before that A_2, B_2 and M are also collinear, yielding the desired result.

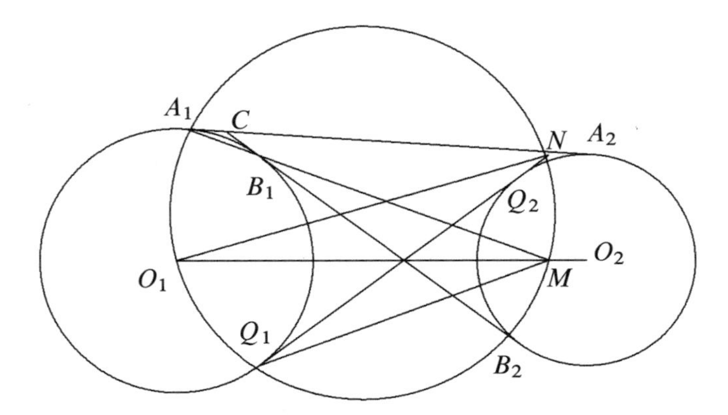

Third Solution Let the centers of the circles be O_1 and O_2. Let the external common tangent be A_1A_2 and the internal common tangent be B_1B_2, intersecting A_1A_2 at C. Now CO_1 and CO_2 are perpendicular since they are bisectors of supplementary angles. Hence A_1B_1 and A_2B_2 are also perpendicular. Let their point of intersection be L. Draw the circles with diameters A_1A_2 and B_1B_2. Then M is one of their points of intersection. Since $O_1A_1 = O_1B_1$ and $O_2A_2 = O_2B_2$, O_1O_2 is the radical axis of the two new circles, and hence passes through M.

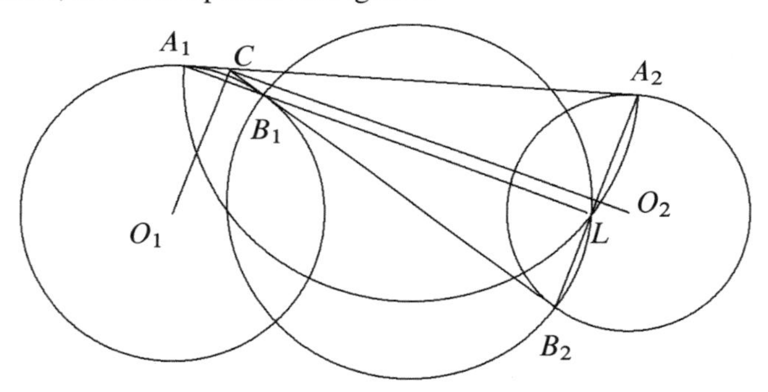

3.9 Problem Set: Geometric Inequalities

Problem 1954.1

In a convex quadrilateral $ABCD$, $AB + BD \leq AC + CD$. Prove that $AB < AC$.

First Solution Let AC meet BD at E. By the Triangle Inequality,

$$AB + CD < AE + BE + CE + DE = AC + BD.$$

Item Price	Total
$40.95	$40.95

$40.95
$40.95
$40.95
$0.00

http://www.amazon.com

For detailed information about this and other orders, please visit Your Account. You can also print invoices, change your e-mail address and payment settings, alter your communication preferences, and much more – 24 hours a day – at http://www.amazon.com/your-account.

Returns Are Easy!

Visit http://www.amazon.com/returns to return any item – including gifts – in unopened or original condition within 30 days for a full refund (other restrictions apply). Please have your order ID ready.

Thanks for shopping at Amazon.com, and please come again!

Your order of February 10, 2011 (Order ID 102–1476795–1543406)

Qty.	Item
	IN THIS SHIPMENT
1	**Hungarian Problem Book IV** Paperback **(** P–1–I30C30 **) 0883858312** **0883858312**

Subtotal
Order Total
Paid via Visa
Balance due

This shipment completes your order.

Have feedback on how we packaged your order? Tell us at www.amazon.com/packaging.

66/DQDwr404R/-1 of 1-//1XSP/second/6097467/0218-15:00/0217-05:14/benjammo   Pack Type: V3

Adding this to the hypothesis $AB+BD \le AC+CD$, we have $2AB < 2AC$ so that $AB < AC$.

Second Solution Let ℓ be the perpendicular bisector of BC. Assume to the contrary that $AB \ge AC$. Then A either lies on ℓ or on the same side of ℓ as C. Since $ABCD$ is a convex quadrilateral, D must lie in the infinite region bounded by AC and the rays BA and BC. Hence D is also on the same side of ℓ as C, so that $BD > CD$. Addition yields $AB + BD > AC + CD$, which contradict the hypothesis.

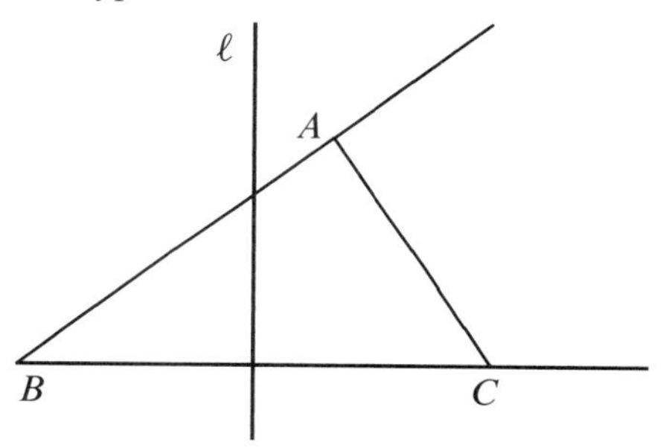

Problem 1955.1

In the quadrilateral $ABCD$, AB is parallel to DC. Prove that if $\angle BAD < \angle ABC$, then $AC > BD$.

First Solution Extend CD to E so that $\angle EAB = \angle ABC$. Then AC and BE intersect at a point F on the perpendicular bisector ℓ of AB. Let BD intersect AC at G. Then G is on CF. Hence it is on the same side of ℓ as B, so that $GA > GB$. Now $\angle GCD = \angle GAB < \angle GBA = \angle GDC$, By the Angle-Side Inequality. so that $GC > GD$ Addition yields $AC > BD$.

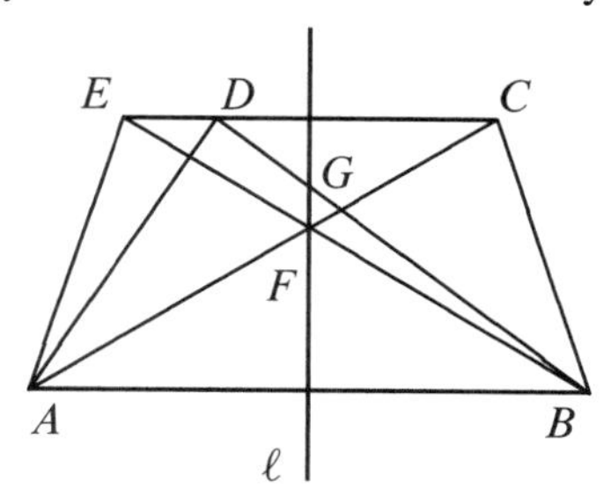

Second Solution Extend CD to E so that $\angle EAB = \angle ABC$. Then B and C are on the same side of the perpendicular bisector ℓ of CE. Hence $BE >$

BC. Since $\angle BDE + \angle BDC = 180°$, one of them is non-acute. It follows from the Angle-Side Inequality that $BD < \max\{BC, BE\} = BE = AC$.

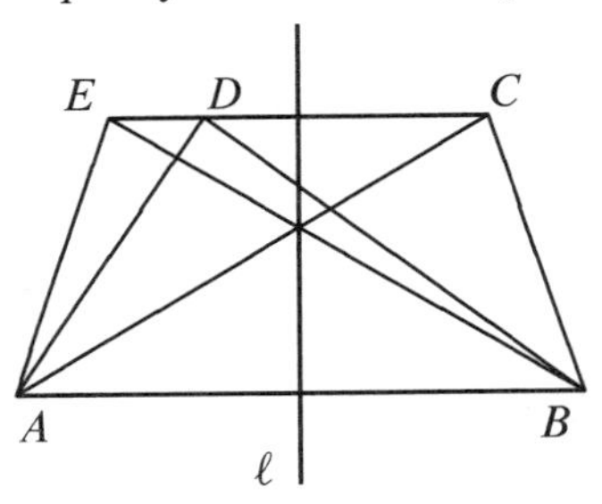

Third Solution Extend CD to E so that $\angle EAB = \angle ABC$. Then we have $\angle BED < \angle AED = \angle BCD < \angle BDE$ by the Exterior Angle Inequality. It follows from the Angle-Side Inequality that $BD < BE = AC$.

Problem 1963.3

Prove that if a triangle is not obtuse, then the sum of the lengths of its medians is at least four times the circumradius of the triangle.

Solution Let m_a, m_b and m_c be the lengths of the medians of triangle ABC which is not obtuse. Let R be its circumradius. Let G be its centroid. Then the circumcenter O lies in one of GAB, GBC and GCA. By symmetry, we may assume that it is GAB. By the Sss Inequality,

$$GA + GB \geq OA + OB$$

so that

$$m_a + m_b \geq 3R.$$

Let F be the midpoint of AB. Then OF is perpendicular to AB, and C is on the same side as O of the perpendicular bisector of OF. Hence $CF \geq CO$ so that $m_c \geq R$. Addition yields the desired result.

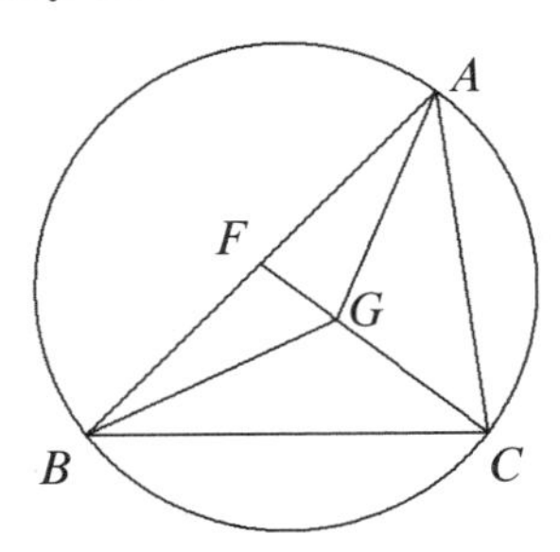

Problem 1952.3

Let $\frac{1}{2} < k < 1$. Let A', B' and C' be points on the sides BC, CA and AB, respectively, of the triangle ABC such that $BA' = kBC$, $CB' = kCA$ and $AC' = kAB$. Prove that the perimeter of the triangle $A'B'C'$ does not exceed k times the perimeter of ABC.

Solution Through A', B' and C', draw lines parallel to AB, BC and CA, cutting CA, AB and BC at E, F and D, respectively. Since $k > \frac{1}{2}$, these points lie on CB', AC' and BA' respectively. Moreover, $DC = kBC$, $EA = kCA$ and $FB = kAB$. Hence triangles AFB', BDC' and CEA' are congruent to one another. By the Triangle Inequality, we have

$$\begin{aligned} B'C' + C'A' + A'B' &\leq B'F + FC' + C'D + DA' + A'E + EB' \\ &= A'C + FC' + B'A + DA' + C'B \\ &= DC + EA + FB \\ &= k(BC + CA + AB). \end{aligned}$$

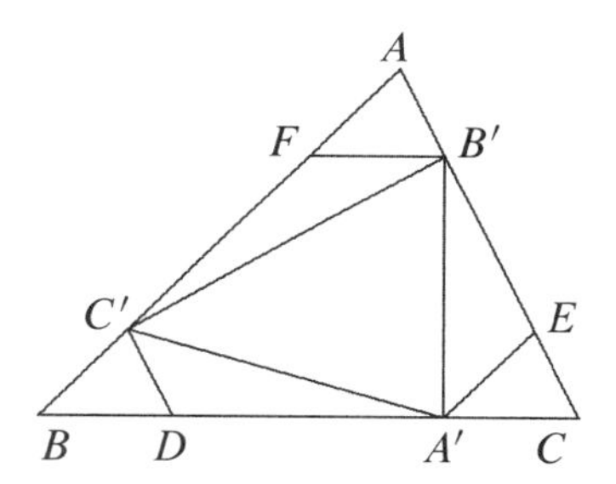

3.10 Problem Set: Combinatorial Geometry

Problem 1961.1

Consider the six distances determined by four points in a plane. Prove that the ratio of the largest of these distances to the smallest cannot be less than $\sqrt{2}$.

First Solution Suppose $\angle BCA \geq 90°$. By Pythagoras' Inequality, we have $AB^2 \geq BC^2 + CA^2$. By symmetry, we may assume that $BC \leq CA$. Then $AB^2 \geq 2BC^2$ so that $AB \geq \sqrt{2}BC$. This includes the degenerate case where C lies on AB. Henceforth, we assume that no three of the points are collinear. Suppose the convex hull of the four points is triangle ABC with D inside. Since $\angle BDC + \angle CDA + \angle ADB = 360°$, at least one of them is non-acute and we have the desired conclusion. Suppose the convex hull is a quadrilateral. Then the sum of its four interior angles is $360°$. Again, at least one of them is non-acute.

Second Solution Let the four points be P, Q, R and S, with PQ the shortest. We take $PQ = 1$. We shall prove that one of the other segments has length at least $\sqrt{2}$. Suppose this is not so. Draw circles ω_1, ω_2, ω_3 and ω_4 with centers P, Q, P and Q and radii 1, 1, $\sqrt{2}$ and $\sqrt{2}$, respectively. Let C and C' be the points of intersection of ω_1 and ω_2, D and D' be those of ω_3 and ω_4, A and A' be those of ω_1 and ω_4, and B and B' be those of ω_2 and ω_3. Now neither R nor S can be inside ω_1 or ω_2 as otherwise the shortest segment will have length less than 1. On the other hand, both must be in the curvilinear quadrilaterals $ACBD$ and $A'C'B'D'$, minus the arcs AD, BD, $A'D'$ and $B'D'$. Otherwise the longest segment will have length at least $\sqrt{2}$.

Note that the quadrilateral $ACBD$ is inside the square $AEBF$ where E is

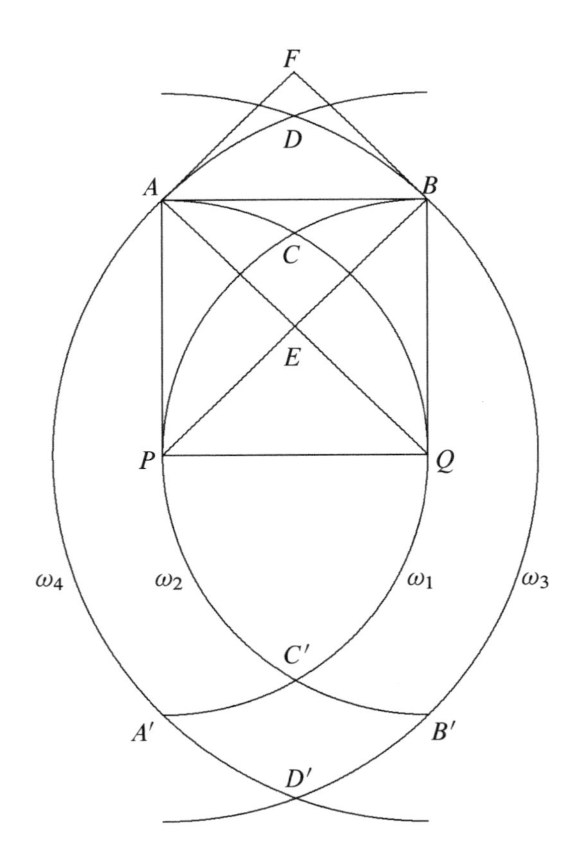

the point of intersection of AQ and BP. If both R and S are inside $ACBD$, then $RS < AB = 1$. Similarly, they cannot both be inside $A'C'B'D'$. Hence R is in one and S the other. However, $RS \geq CC' = \sqrt{3} > \sqrt{2}$, which contradicts our assumption.

Problem 1947.3

The radius of each small disc is half that of the large disc. How many small discs are needed to cover the large disc completely?

Solution We first show that seven such discs are sufficient. Six of them have as diameters the sides of the regular hexagon inscribed in the circumference of the large disc. The seventh one is concentric at O with the large disc. Let AB be one side of that hexagon. It is sufficient to show that every point in the sector OAB of the large disc is covered either by the disc with center O or the disc with center P, the midpoint of AB. Clearly, every point in the segment AB of the large disc is covered by the disc with center P. Let Q and R be the midpoints of OB and OA respectively. Every point inside the equilateral triangle OQR is covered by the disc with center O, while every point inside the quadrilateral $ABQR$ is covered by the disc with center P. We now show that seven such discs are necessary. One of them must cover the point O, and it can have at most one point in common with the circumference of the large disc. Since each small disc can covered at most $\frac{1}{6}$ of the circumference of the large disc, at least six more are required.

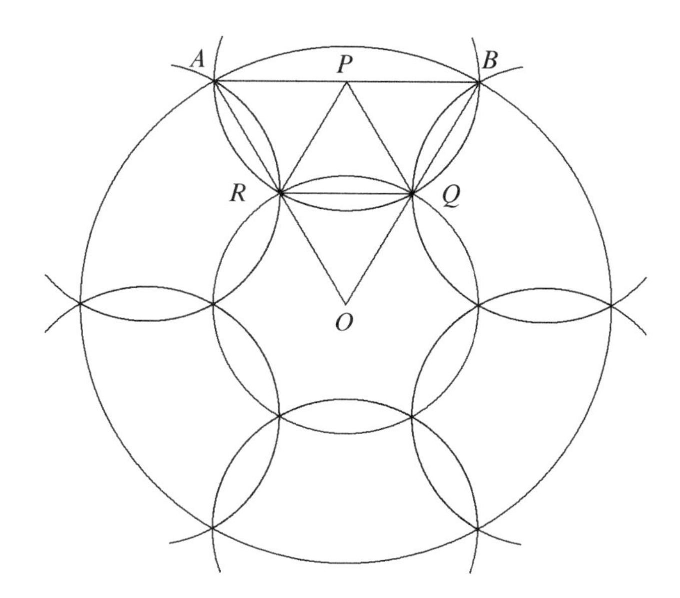

Problem 1958.1

Six points are given on the plane, no three collinear. Prove that three of these points determine a triangle with an interior angle not less than $120°$.

Solution Suppose the convex hull of the six points is a convex hexagon. Since the sum of its six interior angles is $720°$, at least one of them is no less than the average by the Mean Value Principle. Then the vertex of this angle and its two neighbors determine a triangle with an interior angle not less than $120°$. Suppose the convex hull is not a hexagon, so that at least one of the points is inside. Triangulate the convex hull by diagonals from a vertex A. Then some point D inside must lie within some triangle ABC. Since $\angle ADB + \angle BDC + \angle CDA = 360°$, one of them is no less than $120°$.

Problem 1951.3

A plane can be covered completely by four half-planes. Prove that three of these four half-planes are sufficient for covering the plane completely.

First Solution Let Π_1, Π_2, Π_3 and Π_4 be the half-planes. If the result is not true, then there exist four points P_1, P_2, P_3 and P_4 such that P_i is covered only by Π_i for $1 \leq i \leq 4$. Suppose P_1 is between P_2 and P_3. Then Π_1 must also cover one of P_2 or P_3. It follows that no three of the four points are collinear. Suppose their convex hull is a triangle, say $P_1P_2P_3$. Let the extension of P_1P_4 cut P_2P_3 at Q. Then Π_4 must also cover P_1 or Q, and if it covers Q, it must also cover P_2 or P_3. Finally, suppose the convex hull is a convex quadrilateral $P_1P_2P_3P_4$. Let R be the point of intersection of P_1P_3 and P_2P_4. Then the half-plane which covers R must also cover P_1 or P_3 as well as P_2 or P_4. In every case, we have a contradiction.

Second Solution Let Π_1, Π_2, Π_3 and Π_4 be the half-planes. If the result is not true, then there exist four points P_1, P_2, P_3 and P_4 such that P_i is covered only by Π_i for $1 \leq i \leq 4$. Let ℓ be the boundary of Π_4. Then P_4 is on one side of ℓ while P_1, P_2 and P_3 are on the other side. We may assume that P_1 is closest to ℓ. Let ℓ' be the line parallel to ℓ but at half the distance from P_1. Then we still have P_4 on one side of ℓ' while the other three points are on the other side. Let Q_i be the point of intersection of P_iP_4 with ℓ' for $1 \leq i \leq 3$. Since Π_2 or Π_3 does not cover P_1 or P_4, neither covers Q_1. Hence only Π_1 covers Q_1. Similarly, only Π_2 covers Q_2 and only Π_3 covers Q_3. This is impossible since Q_1, Q_2 and Q_3 are collinear.

Third Solution We first prove a lemma. If a line is covered by three rays, then it is covered by two of them. All we need to observe is that two of the rays must be going in the same direction, so that one of them is covered entirely by the other, and is therefore redundant. Let Π_1, Π_2, Π_3 and Π_4 be the half-planes. Let ℓ be the boundary of Π_4. It is covered by Π_1, Π_2 and Π_3. Suppose one of them, say Π_1, covers all of ℓ. Then either Π_1 and Π_4 cover the entire plane, or Π_4 is contained entirely within Π_1, and is therefore redundant. Hence we may assume that each of Π_1, Π_2 and Π_3 covers a ray along ℓ. By the lemma, ℓ is covered by exactly two of the rays, say those covered by Π_1 and Π_2. If the boundaries of these two half-planes are parallel, then they cover the entire plane. Suppose they intersect at some point L. If L is on or inside Π_4, then Π_3 is redundant. If L is outside Π_4, then Π_4 is redundant.

3.11 Problem Set: Trigonometry

Problem 1949.1

Prove that $\sin A + \frac{1}{2}\sin 2A + \frac{1}{3}\sin 3A > 0$ if $0° < A < 180°$.

First Solution We have

$$\begin{aligned}\sin 3A &= \sin(2A + A)\\ &= \sin 2A\cos A + \cos 2A\sin A\\ &= (2\sin A\cos A)\cos A + (\cos^2 A - \sin^2 A)\sin A\\ &= 3\sin A\cos^2 A - \sin^3 A.\end{aligned}$$

Hence

$$\begin{aligned}&\sin A + \frac{1}{2}\sin 2A + \frac{1}{3}\sin 3A\\ &\quad= \sin A + \sin A\cos A + \sin A\cos^2 A - \frac{1}{3}\sin^3 A\\ &\quad= \frac{1}{3}\sin A(2 + 3\cos A + 4\cos^2 A)\\ &\quad= \frac{1}{3}\sin A((1+\cos A)^2 + (1+\cos A) + 3\cos^2 A)\\ &\quad> 0\end{aligned}$$

since $0° < A < 180°$.

Second Solution Since $\sin 2A = 2\sin A\cos A$, we have

$$\sin A + \frac{1}{2}\sin 2A = \sin A(1 + \cos A) > 0$$

for $0° < A < 180°$. Within this range, $\sin 3A < 0$ only for $60° < A < 120°$. However, here

$$\sin A + \frac{1}{2}\sin 2A + \frac{1}{3}\sin 3A > \frac{\sqrt{3}}{2} - \frac{1}{2} - \frac{1}{3} > 0.$$

Third Solution On $(0°, 180°)$, $\sin x > 0$ while $\sin 2x < 0$ only on the subinterval $(90°, 180°)$. We claim that the graph of $y_1 = \sin x$ lies above the graph of $-y_2$, where $y_2 = \frac{1}{2}\sin 2x$. Now the graph of y_2 on $[0°, 90°]$ may be obtained from the graph of y_1 on $[0°, 180°]$ by contracting each point halfway towards the origin. Since y_1 is concave on the interval $[0°, 180°]$, the graph of y_1 lies entirely above the graph of y_2 on $[0°, 90°]$. Now the graph of y_1 is symmetric about the line $x = 90°$ while the graph of $-y_2$ on $[90°, 180°]$ is obtained when the graph of y_2 on $[0°, 90°]$ is reflected about the line $x = 90°$. It follows that $\sin x + \frac{1}{2}\sin 2x > 0$ on $[0°, 180°]$. On this interval, $\sin 3x < 0$ only on the subinterval $(60°, 120°)$. However, here we have $\sin x > \sin 60°$ and $\frac{1}{2}\sin 2x > -\frac{1}{2}\sin 60°$. Hence

$$\begin{aligned}\sin x + \frac{1}{2}\sin 2x + \frac{1}{3}\sin 3x &> \frac{1}{2}\sin 60° - \frac{1}{3}\\ &= \frac{\sqrt{3}}{4} - \frac{1}{3}\\ &> 0.\end{aligned}$$

Problem 1963.2

Prove that $(1 + \sec A)(1 + \csc A) > 5$ if A is an acute angle.

First Solution Using $\sin 2A = 2\sin A\cos A$, we have

$$\begin{aligned}(1 + \sec A)(1 + \csc A) &= 1 + \frac{1}{\sin A} + \frac{1}{\cos A} + \frac{2}{\sin 2A}\\ &\geq 1 + 1 + 1 + 2\\ &= 5\end{aligned}$$

since $0° < A < 90°$. Moreover, since $\sin A$ and $\cos A$ cannot be equal to 1 simultaneously, the inequality is strict.

Second Solution Construct triangle ABC with $\angle C = 90°$ and $CD = 1$, where D is the foot of perpendicular from C to AB. Then $AB > 2CD = 2$ since AB is a diameter of the circumcircle of ABC while CD is one half of a chord of this circle. By the Triangle Inequality, $BC + CA + AB >$

$2AB = 4$. Now $CA = \csc A$ and $AB = CA \sec A = \sec A \csc A$. Since $\angle BCD = \angle A$, $BC = \sec A$. It follows that

$$(1 + \sec A)(1 + \csc A) = 1 + \sec A + \csc A + \sec A \csc A > 5.$$

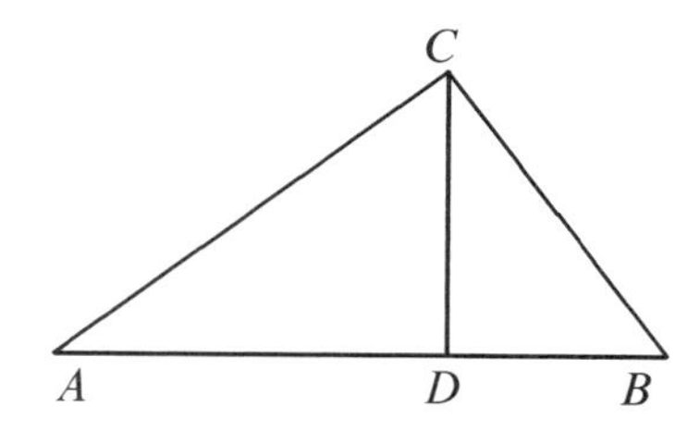

Third Solution We prove that $(1 + \sec A)(1 + \csc A) \geq 3 + 2\sqrt{2}$, with equality if and only if $\angle A = 45°$. We may assume by symmetry that $\angle A < 45°$. Construct triangle ABC with $\angle C = 90°$ and $CD = 1$, where D is the foot of perpendicular from C to AB. Then $AB > 2CD = 2$ since AB is a diameter of the circumcircle of ABC while CD is one half of a chord of this circle. Equality holds if and only if $\angle A = 45°$. Note that $(AC + BC)^2 = AB^2 + 2AC \cdot BC = AB^2 + 2AB \cdot CD \geq 8$ so that $AC + BC \geq 2\sqrt{2}$. Note that we have $CA = \csc A$ and $AB = CA \sec A = \sec A \csc A$. Since $\angle BCD = \angle A$, $BC = \sec A$. It follows that

$$\begin{aligned}(1 + \sec A)(1 + \csc A) &= 1 + \sec A + \csc A + \sec A \csc A \\ &\geq 3 + 2\sqrt{2},\end{aligned}$$

with equality if and only if $\angle A = 45°$.

Fourth Solution We prove that $(1 + \sec A)(1 + \csc A) \geq 3 + 2\sqrt{2}$, with equality if and only if $\angle A = 45°$. We may assume by symmetry that $\angle A < 45°$. Construct triangle ABC with $\angle C = 90°$ and $CD = 1$, where D is the foot of perpendicular from C to AB. Rotate triangle BCD $90°$ about C so that D goes into D' and B goes into B'. Since $\angle BCD = \angle A$, B' lies on the extension of AC. The desired result follows if we can prove that $AB' > XY$, where X on AD and Y on the extension of $D'B'$ are such that C is the midpoint of XY. Let the line through X parallel to BT cut AB' at Z. Then triangles $CB'Y$ and CZX are congruent. Hence $BT = XZ$. Complete the rectangle $AXZW$. Then $AWYB'$ is a parallelogram. Note that in triangle WXY, we have $\angle WXY = 135° - \angle A > 90°$. Hence it is the largest angle of the triangle, so that $AB' = WY > XY$.

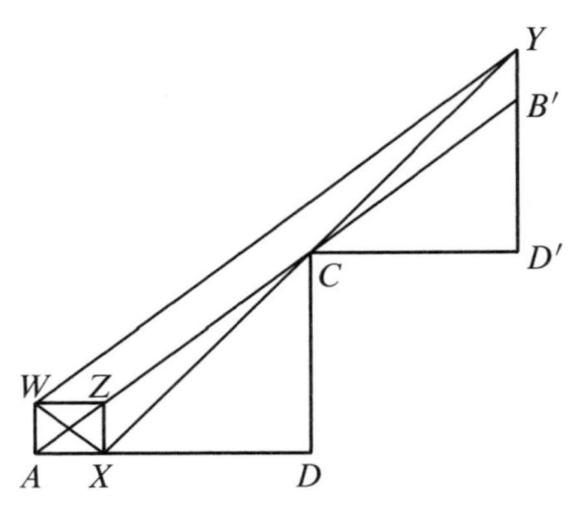

Fifth Solution We prove that $(1 + \sec A)(1 + \csc A) \geq 3 + 2\sqrt{2}$, with equality if and only if $A = 45°$. Since $\sec A \csc A = \frac{2}{\sin 2A} \geq 2$, all we need to prove is $\sec A + \csc A \geq 2\sqrt{2}$. By the Arithmetic-Geometric Mean Inequality, $\sec A + \csc A \geq 2\sqrt{\sec A \csc A}$. By the Root-Mean-Square Geometric Mean Inequality, we have $\sqrt{\sin A \cos A} \leq \sqrt{\frac{\sin^2 A + \cos^2 A}{2}} = \frac{1}{\sqrt{2}}$. It now follows that $\sec A + \csc A \geq \frac{2}{\sqrt{\sin A \cos A}} = 2\sqrt{2}$, with equality if and only if $A = 45°$.

Sixth Solution Let

$$\begin{aligned} f(A) &= (1 + \sec A)(1 + \csc A) \\ &= 1 + \sec A + \csc A + 2\csc 2A. \end{aligned}$$

On the interval $(0°, 90°)$, it is a sum of four convex functions, and is therefore convex. Note that $f(90° - A) = f(A)$. By Jensen's Inequality,

$$\begin{aligned} f(A) &= \frac{f(A) + f(90° - A)}{2} \\ &\geq f(\frac{A + (90° - A)}{2}) \\ &= f(45°) \\ &= 3 + 2\sqrt{2}. \end{aligned}$$

Equality holds if and only if $A = 45°$.

Problem 1953.3

In the convex equilateral hexagon $ABCDEF$, the sum of the interior angles at A, C and E is equal to the sum of the interior angles at B, D and F. Prove that opposite angles of the hexagon are equal.

First Solution We first prove a lemma. In triangle ABC and DEF, if

$$\frac{\sin A}{\sin D} = \frac{\sin B}{\sin E} = \frac{\sin C}{\sin F},$$

then $\angle A = \angle D$, $\angle B = \angle E$ and $\angle C = \angle F$. Applying the Sine Laws to both triangles, we have $\frac{BC}{\sin A} = \frac{CA}{\sin B} = \frac{AB}{\sin C}$ and $\frac{EF}{\sin D} = \frac{FD}{\sin E} = \frac{DE}{\sin F}$. Hence

$$\frac{BC}{EF} = \frac{CA}{FD} = \frac{AB}{DE},$$

so that triangles ABC and DEF are similar, and corresponding angles are therefore equal.

In our hexagon, let $\angle A = 2\alpha$, $\angle C = 2\gamma$ and $\angle C = 2\epsilon$. Since $\angle A + \angle C + \angle E = \angle B + \angle D + \angle F$ and the sum of all six angles is 720°, $\alpha + \gamma + \epsilon = 180°$. Let the length of each side of the hexagon be 1. Then $FB = 2\sin\alpha$, $BD = 2\sin\gamma$ and $DF = 2\sin\epsilon$. Since

$$\frac{FB}{\sin BDF} = \frac{BD}{\sin DFB} = \frac{DF}{\sin FBD},$$

we have

$$\frac{\sin\alpha}{\sin BDF} = \frac{\sin\gamma}{\sin DFB} = \frac{\sin\epsilon}{\sin FBD}.$$

It follows from the lemma that $\angle BDF = \alpha$, $\angle DFB = \gamma$ and $\angle FBD = \epsilon$. Now

$$\begin{aligned}
\angle D &= \angle CDB + \angle BDF + \angle FDE \\
&= \frac{1}{2}(180° - 2\gamma) + \alpha + \frac{1}{2}(180° - 2\epsilon) \\
&= \alpha + 180° - \gamma - \epsilon \\
&= 2\alpha \\
&= \angle A.
\end{aligned}$$

Similarly, $\angle F = \angle C$ and $\angle B = \angle E$.

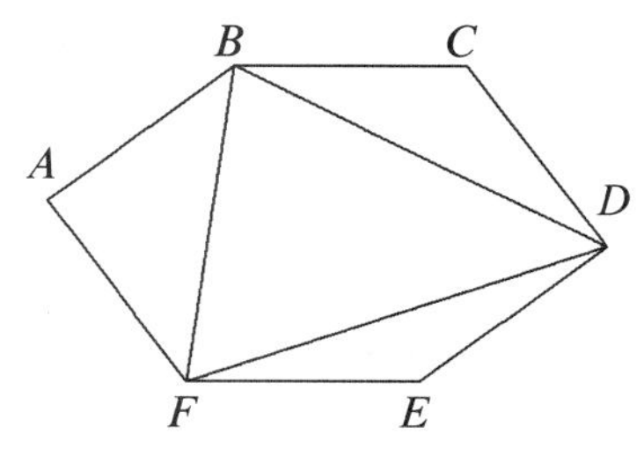

Second Solution Instead of an equilateral hexagon, we only assume that opposite sides are equal. Since the sum of the angles of a hexagon is 720° and $\angle A + \angle C + \angle E = \angle B + \angle D + \angle F$, each side is equal to 360°. Copy triangle FAB to XWZ and triangle BCD to YWX. This is possible since $FA = CD$. We have $WY = BC = EF$, $WZ = DE = AB$ and

$$\begin{aligned}\angle ZWY &= 360^\circ - \angle XWZ - \angle YWX \\ &= 360^\circ - \angle A - \angle C \\ &= \angle E.\end{aligned}$$

Hence we can copy triangle CDE to ZWY. Triangle XYZ can be copied to BDF since corresponding sides are equal. This dissects $ABCDEF$ into three parallelograms $WFAB$, $WBCD$ and $WDEF$. It follows that opposite sides are parallel, and opposite angles are equal.

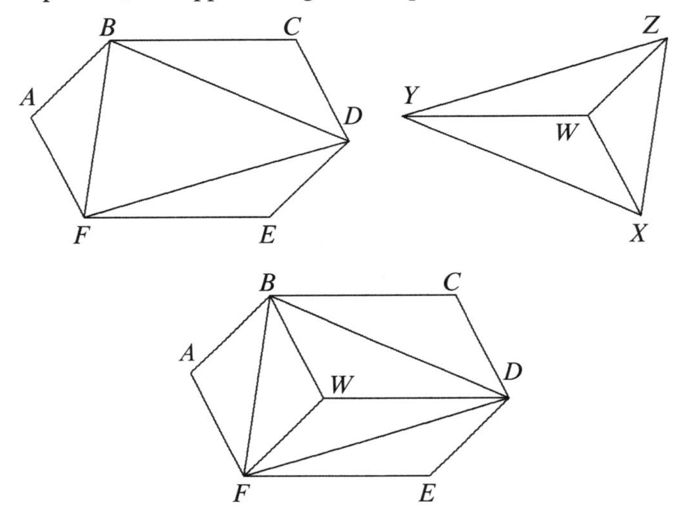

Third Solution Instead of an equilateral hexagon, we only assume that opposite sides are equal. Since the sum of the angles of a hexagon is 720° and $\angle A + \angle C + \angle E = \angle B + \angle D + \angle F$, each side is equal to 360°. Copy $ABCDEF$ to $A_1B_1XWZF_1$. We can make a second copy to $XB_2C_2D_2YW$ since we have $FA = WX$. Now $AB = ZW$, $BC = WY$ and

$$\begin{aligned}\angle ZWY &= 360^\circ - \angle XWZ - \angle YWX \\ &= 360^\circ - \angle D - \angle F \\ &= \angle B.\end{aligned}$$

It follows that we can make a third copy of $ABCDEF$ to $ZWYD_3E_3F_3$. Now the hexagons are congruent to one another. Hence $ZX = YC_2$ and

$ZY = XC_2$, so that C_2XZY is a parallelogram. Similarly, so is A_1XYZ. Hence A_1, X and C_2 are collinear. Since $\angle F_1A_1X = \angle WXC_2$, A_1F_1 is parallel to XW. We can prove in a similar manner that the other pairs of opposite sides are also parallel, and it follows that opposite angles are equal.

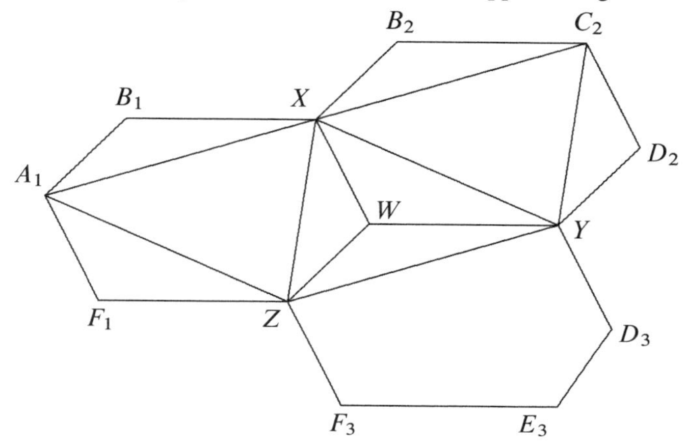

Fourth Solution Instead of an equilateral hexagon, we only assume that opposite sides are equal. Since the sum of the angles of a hexagon is 720° and $\angle A + \angle C + \angle E = \angle B + \angle D + \angle F$, each side is equal to 360°. Let $ABCDGH$ be a centrally symmetric hexagon. Then hypothesis is satisfied and opposite angles are equal. All that remains is show that G coincides with E and H with F. Suppose G does not coincide with E. Since $DG = AB = DE$, $\angle ADE \neq \angle ADG$. We may assume that the latter is larger, as shown in the distorted diagram above. Applying the SaS Inequality to triangles ADE and ADG, we have $AE < AG$. Applying the converse of this inequality to triangles EFA and GHA, we have $\angle EFA < \angle GHA$. Hence

$$\angle ABC + \angle CDE + \angle EFA < \angle ABC + \angle CDG + \angle GHA,$$

but this is a contradiction since both sides are equal to 360°. Hence G must coincide with E and H with F.

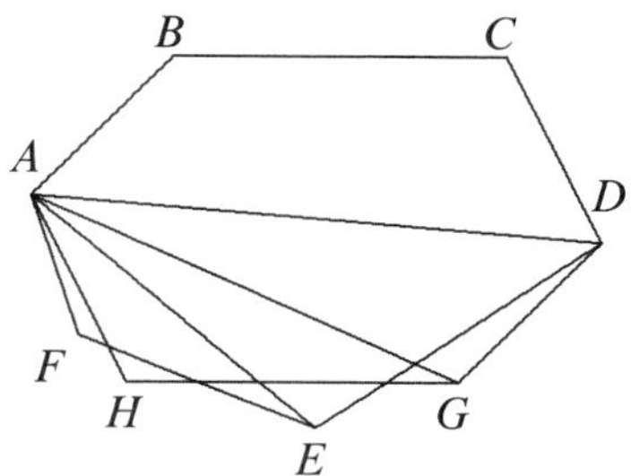

Fifth Solution We first establish a lemma. In the convex quadrilaterals $PQRS$ and $WXYZ$, let $PQ = WX$, $QR = YX$, $RS = YZ$ and $SP = ZW$. If $\angle P < \angle W$, then $\angle R < \angle Y$. To prove this, apply the SaS Inequality to triangles SPQ and ZWX. Then $SQ < ZX$. The desired result follows when we apply the converse of this inequality to triangles QRS and XYZ.

We now return to the problem on hand. Instead of an equilateral hexagon, we only assume that opposite sides are equal. Suppose $\angle A = \angle D$. Then triangles FAB and CDE are congruent, so that $BF = CE$. Hence triangles CBE and FEB are also congruent. It follows easily that $\angle B = \angle E$ and $\angle C = \angle F$. This means that if one of the three equality holds, then all three hold. Suppose one of them does not hold. Then none of them holds, and two of the inequalities must go the same way. By symmetry, we may assume that $\angle A < \angle D$ and $\angle C < \angle F$. Applying the lemma to the quadrilaterals $ABEF$ and $DEBC$, we have $\angle FEB < \angle CBE$ and $\angle DEB < \angle ABE$. Addition yields $\angle E < \angle B$. This is a contradiction.

Sixth Solution Instead of an equilateral hexagon, we only assume that opposite sides are equal. Since the sum of the angles of a hexagon is 720° and $\angle A + \angle C + \angle E = \angle B + \angle D + \angle F$, each side is equal to 360°. Suppose $\angle A \neq \angle D$. We may assume that $\angle A < \angle D$. Then $\angle ABF + \angle AFB > \angle DEC + \angle DCE$. Applying the SaS Inequality to triangles FAB and CDE, we have $BF < CE$. Applying the converse of this inequality to triangles BCE and EFB, we have $\angle CBE > \angle FEB$. We also have $\angle EBF + \angle EFB > \angle BCE + \angle EC$. Addition yields $\angle B + \angle D + \angle F > \angle A + \angle C + \angle E$, a contradiction.

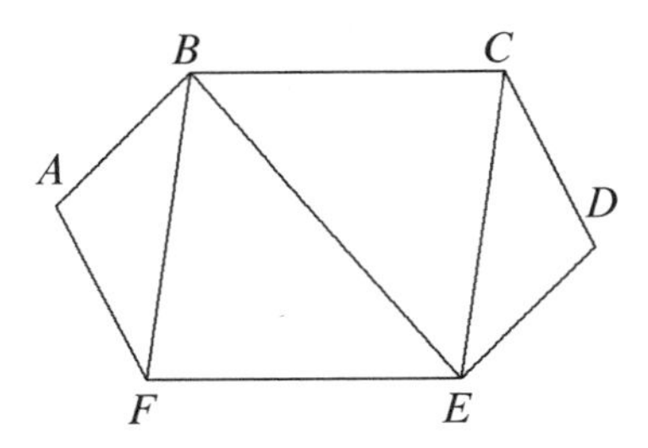

Problem 1959.2

A vertical pole stands on a horizontal plane. The distances from the base of the pole to three other points on the plane are 100, 200 and 300 metres respectively. The sum of the angles of elevation from these three points to the top of the pole is 90° . What is the height of the pole?

First Solution Let the height of the pole be x meters, and the angles of elevation from the points 100, 200 and 300 meters away be α, β and γ respectively. Then

$$\begin{aligned}\tan\gamma &= \tan(90^\circ - (\alpha+\beta)) \\ &= \cot(\alpha+\beta) \\ &= \frac{1}{\tan(\alpha+\beta)} \\ &= \frac{1-\tan\alpha\tan\beta}{\tan\alpha+\tan\beta}.\end{aligned}$$

It follows that

$$\frac{x}{300} = \frac{1-\frac{x^2}{20000}}{\frac{x}{100}+\frac{x}{200}} = \frac{20000-x^2}{300x}.$$

From $x^2 = 20000 - x^2$, we have $x = 100$, the negative root being extraneous.

Second Solution Choose four point A, B, C and D on a straight line such that $AB = BC = CD = 100$. Let E and F be points on the opposite sides of AD such that EB and FD are perpendicular to AD, and each is the length of the pole. Then $EA = EC$. Since $\angle EAB + \angle EDB + \angle FAD = 90^\circ$ by the hypothesis, $\angle EAF + \angle EDF = 180^\circ$, so that $AEDF$ is a cyclic quadrilateral. We now have $\angle AEF = \angle ADF = 90^\circ$, so that EAC is an isosceles right triangle. It follows that $\angle EAC = 45^\circ$ and $EB = AB = 100$.

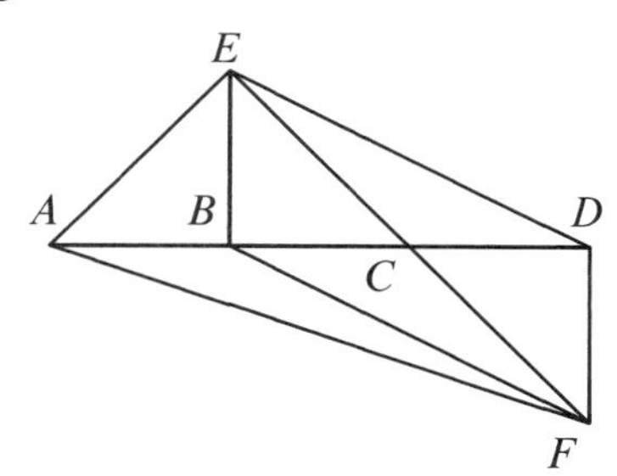

Third Solution Let the angles of elevation from the points 100, 200 and 300 meters away be α, β and γ respectively. We construct three triangles $D_1C_1B_1$, $D_2C_2A_2$ and $D_3A_3B_3$ such that $B_1C_1 = 500$, $C_2A_2 = 400$ and $A_3B_3 = 300$. Moreover, $\angle D_2A_2C_2 = \alpha = \angle D_3A_3B_3$, $\angle D_1B_1C_1 = \beta = \angle D_3B_3A_3$ and $\angle D_1C_1B_1 = \gamma = \angle D_2C_2A_2$. Then the distance from D_i to the opposite side, $1 \le i \le 3$, is the height of the pole. We now have $\angle B_1D_1C_1 = 90^\circ+\alpha$, $\angle C_2D_2A_2 = 90^\circ+\beta$ and $\angle A_3D_3B_3 = 90^\circ+\gamma$. The sum of these three angles is 360°. Hence the three triangles can fit together to form a triangle ABC at its incenter D. Since $BC^2 = AB^2 + BC^2$, we have $\angle CAB = 90°$ by the converse of Pythagoras' Theorem. Hence $\alpha = 45°$ and the height of the pole is 100 meters.

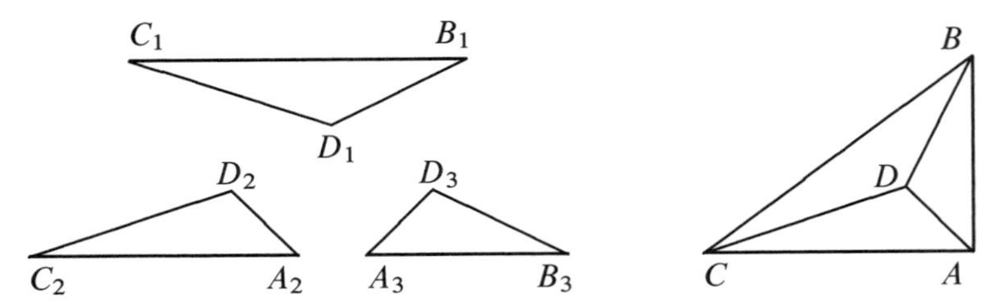

3.12 Problem Set: Solid Geometry

Problem 1948.2

Prove that except for any tetrahedron, no convex polyhedron has the property that every two vertices are connected by an edge. Degenerate polyhedra are not considered.

First Solution Suppose some convex polyhedron with at least five vertices has the property that every two vertices are connected by an edge. Then every face of it must be a triangle, as otherwise there are two vertices on a non-triangular face which are not connected by an edge. Let AB be an edge. Then it is an edge of two triangular faces. Since there are at least five vertices, there exists a vertex C not on either of these faces. Nevertheless, AC and BC are edges of the convex polyhedron. If we cut along the plane of ABC, we can divide the original polyhedron into two pieces each of which is a convex polyhedron. Each piece must have a fourth vertex other than A, B and C. However, such a vertex in one piece cannot be connected by an edge to such a vertex in the other piece. This is a contradiction.

Second Solution Suppose some convex polyhedron with $n \geq 5$ vertices has the property that every two vertices are connected by an edge. Then there are exactly $\binom{n}{2}$ edges. Since each face has at least 3 edges, the number of faces is at most $\frac{2}{3}\binom{n}{2}$. From Euler's Formula, we have $n + \frac{2}{3}\binom{n}{2} \geq \binom{n}{2} + 2$. This simplifies to $(n-3)(n-4) \leq 0$, which contradicts $n \geq 5$.

Problem 1962.3

Let P be a point in or on a tetrahedron $ABCD$ which does not coincide with D. Prove that at least one of the distances PA, PB and PC is shorter than at least one of the distances DA, DB and DC.

First Solution We may take D to be vertically above P, and consider the horizontal plane through P. Since P is in $ABCD$, not all of A, B or C are above this plane. Suppose A is on or below this plane. Then $\angle DPA \geq 90° > \angle PDA$. Hence $DA > PA$ by the Angle-Side Inequality.

Second Solution We may take ABC to be horizontal and D above it. The vertical line through P cuts this horizontal plane at Q and one of the lateral faces at R. We may assume that R is in BAD. The line through R perpendicular to AB cuts AB at S and one of the lateral edges at T. We may assume that T is on DA. Now $\angle APR > \angle AQR = 90° > \angle AQP$ and $\angle ART > \angle AST = 90° > \angle ATR$. By the Angle-Side Inequality, $AP < AR < AT < AD$.

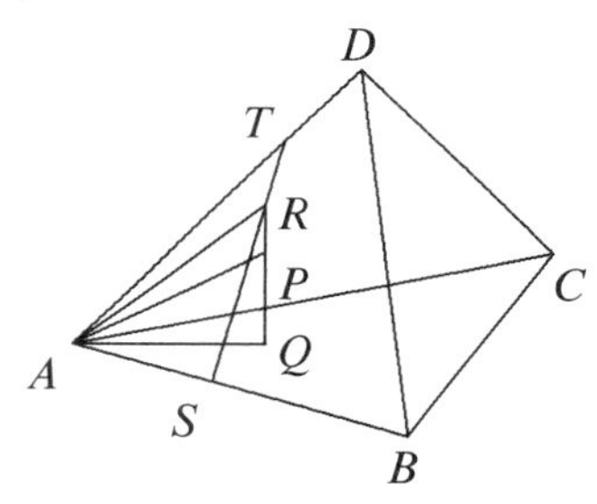

Third Solution We may take D to be vertically above P, and consider the horizontal plane through the midpoint of DP. This divide space into two half-spaces, consisting of points closer to D than to P, and points closer to P than to D, respectively. Since P is in $ABCD$, some of A, B and C must be in the same half-space as P. Suppose it is A. Then $DA > PA$.

Problem 1957.1

Let ABC be an acute triangle. Consider the set of all tetrahedra with ABC as base such that all lateral faces are acute triangles. Find the locus of the projection onto the plane of ABC of the vertex of the tetrahedron which ranges over the above set.

Solution We may take ABC to be horizontal and the fourth vertex V above this plane. Since ABC is acute, its circumcenter is inside, and $AFBDCE$ is a convex hexagon where AD, BE and CF are diameters of the circumcircle. We claim that the locus of the projection of V onto the plane of ABC is the interior of this hexagon. Consider first any point P in the interior of this hexagon. Then it lies between the lines BF and CE which are perpendicular to BC. Hence the vertical line through P lies between the vertical planes through B and C perpendicular to BC. For any point V on this line, both

$\angle VBC$ and VCB are acute, and if $VP > \frac{1}{2}BC$, then $\angle BVC$ is also acute. Using the same argument on the other two sides of triangle ABC, we can show that if V is vertically above P and $VP > \frac{1}{2}\max\{BC, CA, AB\}$, then all of triangles VBC, VCA and VAB are acute. Now let V be a point above the plane of ABC such that all of triangles VBC, VCA and VAB are acute. Since $\angle VBC$, V is on the same side as C of the vertical plane through B perpendicular to BC. Since $\angle VCB$ is acute, V is on the same side as B of the vertical plane through C perpendicular to BC. It follows that the projection P of V onto the plane of ABC lies between the lines BF and CE which are perpendicular to BC. Similarly, P lies between CD and AF, as well as between AE and BD. It follows that P indeed lies in the interior of $AFBDCE$.

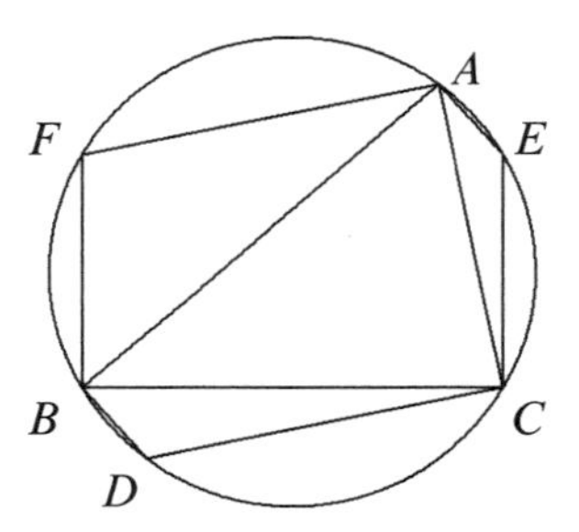

Problem 1954.2

Prove that if every planar section of a three-dimensional solid is a circular disk, then the solid is a sphere.

First Solution Consider a horizontal planar section of the solid. Its boundary is a circle ω with center C. The vertical line through C cuts the solid at a point A above the plane of ω and a point B below it. Consider now a vertical planar section of the solid containing AB. Its boundary is a circle ω' which cuts ω at P and Q. Since AB is perpendicular to the chord PQ of ω' and passes through its midpoint C, it is a diameter of ω'. Since the vertical planar section is chosen arbitrarily, the solid is a sphere with diameter AB.

Second Solution Consider a horizontal planar section of the solid. Its boundary is a circle ω. From any point Q inside ω. draw a ray above the plane of ω, meeting the boundary of the solid at P. Let Ω be the sphere with ω on its boundary and passing through P. Consider now any plane containing PQ. It will intersect ω at two points A and B. The intersection of this plane with either the given solid or the sphere Ω is the same circle, namely, the circumcircle of triangle PAB. It follows that the solid coincides with Ω, and is therefore a sphere.

4

Looking Back

In this chapter, we look back at some of the problems and their solutions, in both this volume and in Hungarian Problem Book III. We look at various extensions and generalizations as well as related results. While this material is not needed for solving the problems at hand, they may provide valuable insight in solving other problems.

4.1 Discussion on Combinatorics

Most of our discussion in this section is on graph theory. We begin by looking back at the following problem which has become a classic.

Problem 1947.2

Prove that in any group of six people, either there are three people who know one another or three people who do not know one another. Assume that "knowing" is a symmetric relation.

This problem can be rephrased in graph theoretic terms in at least two ways other than that used in its solution. In the first version, which has a symmetric form, we have a graph on six vertices. We wish to conclude that either the graph or its complement contains a complete subgraph on 3 vertices.

This can be generalized as follows. What is the minimum number $m(k)$ such that for any graph with $m(k)$ vertices, either the graph or its complement contains a complete subgraph on k vertices?

Clearly, $m(1) = 1$ and $m(2) = 2$. The result of Problem 1947.2 shows that $m(3) \leq 6$. In fact, $m(3) = 6$ since neither a pentagon nor its complement contains a triangle.

In general, $m(k) \leq \binom{2k-2}{k-1}$. We will justify this result later, but it is far from being the best. On the other hand, Erdős proved that $m(k) > 2^{\frac{k}{2}}$ using powerful existence arguments. It may be interesting to find constructive proofs.

Our problem is meaningful even for graphs with infinitely many vertices. We claim that either such a graph or its complement contains an infinite complete subgraph.

Pick an arbitrary vertex A_1. If there are vertices not joined to A_1, pick an arbitrary one A_2. After A_k has been picked, if there are vertices not joined to any of A_1, A_2, ..., A_k, pick an arbitrary one A_{k+1}. Suppose we can continue forever. Then the complement of the graph has an infinite complete subgraph. Hence we assume that there is a last vertex A_n which can be picked.

Consider all remaining vertices. Each of them is joined to at least one of these n vertices. By the Finite Union Principle, infinitely many of them are joined to one of them. Rename it B_1, and consider only vertices joined to it. Note that other A's are no longer under consideration. Thus we can pick a new arbitrary vertex A_1, and repeat the process above. Either the complement has an infinite complete subgraph, or we can pick a vertex B_2 which is joined to infinitely many vertices. Continuing this way, we either show that the complement has an infinite complete subgraph, or construct an infinitely complete subgraph of the graph itself.

In the second alternative version of Problem 1947.2, we wish to conclude that any graph on six vertices, no three independent, contains a complete subgraph on 3 vertices.

This can be generalized as follows. What is the minimum number $n(k)$ such that any graph with $n(k)$ vertices, no three independent, contains a complete subgraph on k vertices?

We have $n(1) = 1$ because the graph with a single vertex does not have 3 independent vertices and contains a complete subgraph on 1 vertex. We next show that $n(2) = 3$. If the graph has 3 vertices, and no 3 are independent, then 2 of them must be joined by an edge. This constitutes a complete subgraph on 2 vertices. On the other hand, the graph consisting of two isolated vertices does not have 3 independent vertices, and yet contains no complete subgraph on 2 vertices. The result of Problem 1947.2 shows that $n(3) \leq 6$, and the pentagon shows that $n(3) = 6$.

In general, consider a graph with no 3 independent vertices and does not contain a complete subgraph on k vertices. Pick any vertex A. Consider those vertices not joined to A. They must form a complete subgraph. If 2 of them are not joined, they will form a set of 3 independent vertices with A. Hence there are at most $k-1$ of them. Consider now those vertices joined to A. They cannot contain a complete subgraph on $k-1$ vertices, as otherwise that will become a complete subgraph on k vertices when we throw in A. Hence there are at most $n(k-1)-1$ of them. Thus the total number of vertices is at most $1+(k-1)+n(k-1)-1 = n(k-1)+k-1$. If we add one more vertex, we will have a subgraph on k vertices. Hence $n(k) \leq n(k-1)+k$.

Iterating this inequality, we have

$$n(k) \leq 1+2+\cdots+k = \frac{k(k+1)}{2}.$$

This is a quadratic upper bound. The idea for a linear lower bound $n(k) \geq 3k-3$ comes unexpectedly from the second solution to the following problem.

Problem 1952.2

Let n be an integer greater than 1. From the integers from 1 to $3n$, $n+2$ of them are chosen arbitrarily. Prove that among the chosen numbers, there exist two of them whose difference is strictly between n and $2n$.

Place the $3k-3$ vertices of the graph around a circle. Two vertices are joined by an edge if and only if there are at most $n-2$ other vertices between them along the circle. Then any n vertices in a continuous block form a complete subgraph. At the same time, the graph cannot have three independent vertices as every two of them will have at least $n-1$ other vertices between them.

We now generalize the graph theoretic formulation of Problem 1947.2 used in the solutions to the case where the edges of a complete graph are painted in n colors. Let $k_1, k_2, \ldots, k_n$ be positive integers. It follows from a profound result known as **Ramsey's Theorem** that if the graph is large enough, there exists an index i such that the graph contains a complete subgraph on k_i vertices, all edges of which are painted in the ith color.

We denote the minimum number of vertices of such a graph by $R(n; k_1, k_2, \ldots, k_n)$. Then $m(k) = R(2; k, k)$ while $n(k) = R(2; 3, k)$. We have $R(2; k, 1) = R(2; 1, \ell) = 1$, $R(2; k, 2) = k$ and $R(2; 2, \ell) = \ell$. A most useful result is $R(2; k, \ell) \leq R(2; k-1, \ell) + R(2; k, \ell-1)$.

Let the colors be red and blue. Consider a graph with $R(2; k-1, \ell) + R(2; k, \ell-1)$ vertices. Pick any vertex A. It is either joined to at least $R(2; k-1, \ell)$ vertices by red edges, or joined to at least $R(2; k, \ell-1)$ vertices by blue edges. One of these two cases must occur since there are only $R(2; k-1, \ell) + R(2; k, \ell-1) - 1$ other than A. By symmetry, we may assume that it is the first case. Among the $R(2 : k-1, \ell)$ vertices joined to

A by red edges, we either have a complete red subgraph on $k - 1$ vertices, or a complete blue subgraph on ℓ vertices. In the former case, we have a complete red subgraph on k vertices when we throw in A. In the latter case, we already have a complete blue subgraph on ℓ vertices.

We are now ready to justify that $m(k) \le \binom{2k-2}{k-1}$ by proving the more general result

$$R(2;k,\ell) \le \binom{k+\ell-2}{k-1} = \binom{k+\ell-2}{\ell-1}.$$

We use induction on $k + \ell$. The base is easy to verify, and

$$\begin{aligned} R(2;k,\ell) &\le R(2;k-1,\ell) + R(2;k,\ell-1) \\ &\le \binom{k+\ell-3}{k-2} + \binom{k+\ell-3}{k-1} \\ &= \binom{k+\ell-1}{k-1}. \end{aligned}$$

Finally, we prove a result with three colors, namely, $R(3;3,3,3) = 17$. The proof of the upper bound follows the first solution to Problem 1947.2. Let the colors be red, yellow and blue. Consider a particular vertex A and the sixteen edges incident with it. By the Mean Value Principle, at least six of them have the same color, say red. Consider the complete subgraph on these six vertices. If any edge is red, we have a red triangle. If none of them is red, then by Problem 1947.2, we must have either a yellow triangle or a blue triangle.

The lower bound is obtained by finding a way of painting the edges of a complete graph on sixteen vertices in three colors so that there is no triangle of any color. Label the vertices with the binary quadruples (0,0,0,0) to (1,1,1,1).

Each edge is labeled with the sum of the labels of its two endpoints, using component-wise addition in modulo 2. Thus the edges are also labeled with binary quadruples, except that (0,0,0,0) cannot appear.

Paint the edge red if its label is (1,1,1,1), (0,0,0,1), (0,0,1,0), (0,1,0,0) or (1,0,0,0). Paint the edge yellow if its label is (1,1,1,0), (0,1,1,1), (1,1,0,0), (0,1,1,0) or (0,0,1,1). Paint the edge blue if its labels is (1,0,1,1), (1,1,0,1), (1,0,0,1), (1,0,1,0) or (0,1,0,1).

In any triangle, the label of each edge is the sum of the labels of the other two edges. In each of the three sets of five labels given above, the sum of any two labels in the set is not in the set. Hence there are no triangles with edges in the same color.

We now turn to another problem.

Problem 1957.2

A factory manufactures several kinds of cloth, using for each of them exactly two of six different colors of silk. Each color appears on at least three kinds of cloth, each with a distinct second color. Prove that there exist three kinds of cloth such that between them, all six colors are represented.

In the third solution, if we superimpose the two sets of three independent edges of the graph G, we obtain the subgraph of G shown in the diagram below.

This subgraph is a cycle which passes through each vertex once and only once, and is called a **Hamiltonian cycle**. It is still an open problem to find a necessary and sufficient condition for a graph to have a Hamiltonian cycle.

The G has six vertices each of degree at least 3. More generally, if a graph has n vertices each of degree at least $\frac{n}{2}$, then the graph has a Hamiltonian cycle. This result is known as **Dirac's Theorem**.

We arrange the vertices $A_1, A_2, \ldots, A_n$ in a circle. If A_i is joined to A_{i+1} for $1 \leq i \leq n-1$ and A_n is also joined to A_1, then we have a Hamiltonian cycle. By symmetry, we may assume that A_n is not joined to A_1. We call this a gap.

We shall show that under the hypothesis of the theorem, we can reduce the number of gaps by at least 1. This way, all gaps will eventually disappear, and we will have a Hamiltonian cycle.

We begin with an illustration using the graph in the diagram below.

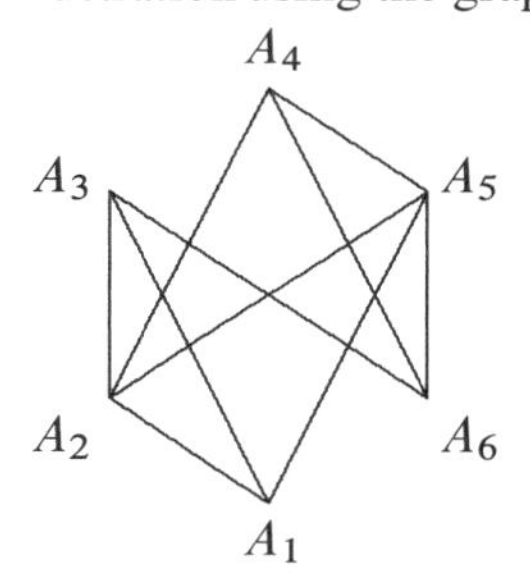

At this moment, A_1A_6 is a gap. We list the vertices incident with either of them in the following chart.

	A_2	A_3	A_4	A_5	A_6
A_1	yes	yes	no	yes	no
	A_1	A_2	A_3	A_4	A_5
A_6	no	no	yes	yes	yes

We have two yeses in the second column from the right. We will interchange A_6 with A_5, replacing A_1A_6 and A_4A_5 by A_1A_5 and A_4A_6. From the chart, neither of the latter is a gap. Whether A_4A_5 is a gap or not, we have already reduced the number of gaps by 1. The graph obtained is shown in the diagram below.

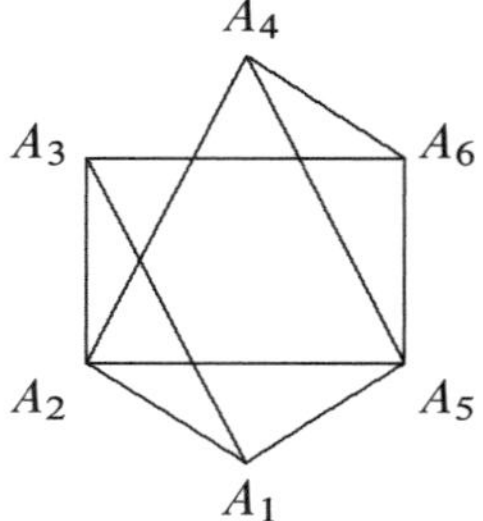

We now relabel $A_4, A_6, A_5, A_1, A_2, A_3$ as $A_1, A_2, A_3, A_4, A_5, A_6$ and apply the same process again. This will remove the last gap.

In general, to remove the gap A_1A_n, we seek a pair of vertices A_k and A_{k+1}, $1 \le k \le n-1$, such that A_k is incident with A_n and A_{k+1} is incident with A_1. The following chart will facilitate the search.

	A_2	A_3	$\cdots$	A_n
A_1	?	?	$\cdots$	no
	A_1	A_2	$\cdots$	A_{n-1}
A_n	no	?	$\cdots$	?

Since each of A_1 and A_n has degree at least $\frac{n}{2}$, the total number of yeses in the two rows is at least n. Since there are $n-1$ columns, the Pigeonhole Principle guarantees that two yeses will be in the same column. This will give us the k we seek. Now reverse the internal order of $A_{k+1}, A_{k+2}, \ldots, A_n$. The net effect is replacing A_1A_n and A_kA_{k+1} by A_1A_{k+1} and A_kA_n, and neither of the latter is a gap. This completes the proof of Dirac's Theorem.

Later, Pósa refined this result as follows. In a graph with n vertices where $n \ge 3$, if for every $k \le \frac{n-1}{2}$, the number of vertices of degree at most k is less than k, then the graph has a Hamiltonian cycle.

Finally, we look back at the following problem from Hungarian Problem Book III.

Problem 1933.2

Sixteen squares of an 8×8 chessboard are chosen so that there are exactly two in each row or column. Prove that eight white pawns and eight black pawns can be placed on these sixteen squares such that there is one white pawn and one black pawn in each row and in each column.

In the Second Solution, we construct a bipartite graph with eight vertices representing the rows and eight vertices representing the columns of the chessboard. An edge joining a vertex on each side represents a given square at the intersection of the row and column represented by the vertices. Thus there are exactly 16 edges, and the graph is regular of degree 2. The conclusion is that there is a set of eight independent edges. Actually, the graph is a disjoint union of two such sets.

In fact, a regular bipartite graph of degree $k > 2$ is the disjoint union of k independent sets of edges. We prove this via a more general result dealing with bipartite graphs which are not necessarily regular.

Let X be the set of vertices on one side and Y the set of vertices on the other side. For any subset S of X, define $N(S)$ as the subset of Y consisting all vertices joined to at least one vertex in S. If $|S| \leq |N(S)|$ for all S, then the graph has $|X|$ independent edges. Note in particular that if we take $S = X$, then $N(X)$ is a subset of Y. It follows from the hypothesis that $|X| \leq |Y|$.

This result is known as **Hall's Theorem**. We first apply it to prove that a regular bipartite graph with $2n$ vertices has n independent edges. Observe first that we must have n vertices in each of the two sides X and Y.

Let k be the degree of each vertex. Consider any subset S of X with m vertices. There are mk edges joining these vertices to the other side. Since the degree of each vertex on the other side is also k, $N(S)$ must contain at least m vertices since some of its vertices may be joined to vertices other than those in S. Hence the hypothesis of Hall's Theorem is satisfied, and the conclusion follows.

Once we have n independent edges, we can remove them without removing any vertices. The remaining graph is still a regular bipartite graph, with degree $k - 1$. Repeating this process, we can decompose the original graph into the disjoint union of k sets of n independent edges.

We now derive Hall's Theorem. A set of vertices in a graph is said to be an **edge cover** if every edge of the graph is incident with at least one vertex in the cover. Clearly, the entire set of vertices is an edge cover. We are interested only in edge covers that are as small as possible. Note that an edge cover cannot be smaller than the maximum number of independent edges in the graph.

It turns out that if the graph is bipartite, then the minimum number of vertices in an edge cover is equal to the maximum number of independent edges. This result is known as **König's Theorem**. We prove this theorem by induction on the number of edges in the bipartite graph G. Let k be the minimum number of vertices in an edge cover. We wish to show that G has at most k independent edges.

The result is trivial if no vertex has degree greater than one. Hence we may assume that there is a vertex A of degree at least 2. Let X and Y be the sets of vertices on the two sides of G. By symmetry, we may assume that A is in X, and is joined to B and C in Y.

There are three cases to consider.

1. The minimum number of vertices in an edge cover of the graph $G-\{\text{AB}\}$ is k.
2. The minimum number of vertices in an edge cover of the graph $G-\{\text{AC}\}$ is k.
3. The graph $G-\{\text{AB}\}$ has an edge cover E with $k-1$ vertices and the graph $G-\{\text{AC}\}$ has an edge cover F with $k-1$ vertices.

If either (1) or (2) holds, we already have k independent edges in a subgraph of G. Hence we will have k independent edges in G itself. We now prove that (3) cannot hold.

Suppose to the contrary that (3) holds. Note that A is not in E or F, as otherwise G will have an edge cover with $k-1$ vertices. Similarly, B is not in E and C is not in F. On the other hand, B must be in F and C must be in E.

Consider the set $(E \cap F \cap X) \cup ((E \cup F) \cap Y)$. We claim that it is an edge cover of G. Consider any edge PQ with P in X and Q in Y. If Q is in E or F, Q is in this set and the edge PQ is covered. Suppose Q is not in E or F. Then Q is not B or C. Since E is an edge cover of $G-\{\text{AB}\}$, P must be in E. Similarly, P must also be in F, and is therefore in this set. Again, the edge PQ is covered.

It may appear that the set $(E \cup F) \cap X) \cup (E \cap Y) \cup (F \cap Y)$ is also an edge cover of G. However, we must add the vertex A in order to cover the edges AB and AC. It follows that

$$\begin{aligned}
2k &\leq |(E \cap F \cap X) \cup ((E \cup F) \cap Y)| \\
&\quad + |\{\text{A}\} \cup ((E \cup F) \cap X) \cup (E \cap F \cap Y)| \\
&= 1 + |E \cap F| + |E \cup F| \\
&= 1 + |E| + |F| \\
&= 2k - 1.
\end{aligned}$$

This contradiction completes the proof of König's Theorem.

We now derive Hall's Theorem from König's Theorem. Let E be any edge cover. Applying the hypothesis of Hall's Theorem to the set $S = X - (E \cap X)$, we have $|S| \leq |N(S)|$. Since E is an edge cover, $N(S)$ is a subset of $E \cap Y$. Hence $|X| = |S| + |E \cap X| \leq |E \cap Y| + |E \cap X| = |E|$. Now X is itself an edge cover. Hence it is an edge cover with the minimum number of vertices. By König's Theorem, the maximum number of independent edges is also $|X|$.

4.2 Discussion on Number Theory

We begin by looking back at the following problem from Hungarian Problem Book III.

Problem 1938.1

Prove that an integer can be expressed as the sum of two squares if and only if so can the number which is twice that integer.

Since $2 = 1 + 1$ is a sum of two squares, one direction of the above problem is a special case of a more general result, namely, if two integers can be expressed as the sum of two squares, then so can their product. This can be justified by the following computation.

$$\begin{aligned}(a^2+b^2)(c^2+d^2) &= a^2c^2 + b^2d^2 + a^2d^2 + b^2c^2 \\ &= a^2c^2 + 2abcd + b^2d^2 \\ &\quad + a^2d^2 - 2abcd + b^2c^2 \\ &= (ac+bd)^2 + (ad-bc)^2.\end{aligned}$$

Since every integer is a product of prime numbers, we can answer the general question about which integers can be expressed as the sum of two squares by investigating which prime numbers have this property. We have already seen that 2 has that property. The odd prime numbers can be classified as being of the form $4k + 1$ or of the form $4k + 3$. These two classes of prime numbers behave very differently with regard to being expressible as the sum of two squares.

Before pursuing this further, we digress and turn our attention to the following problem.

Problem 1951.2

For which positive integers m is $(m-1)!$ divisible by m?

In the solution, we use the fact that if p is a prime, then $(p-1)!$ is not divisible by p. This simple fact also follows from a much stronger result that if p is a prime, then $(p-1)!+1$ is divisible by p. This is known as **Wilson's Theorem.**

We present a proof which makes significant use of geometric intuition. Let $A_0, A_1, \ldots, A_{p-1}$ be p points evenly spaced around a circle. We construct a directed polygonal path starting from A_0, visiting every other point exactly once in an arbitrary order before returning to A_0. Each permutation of $\{1, 2, \ldots, p-1\}$ generates a unique path. The diagram below on the left illustrates the case $p = 7$ with the permutation $\langle 2, 1, 6, 4, 3, 5\rangle$. Hence the total number of paths in general is $(p-1)!$. We want to show that when this number is divided by p, the remainder is $p-1$. Equivalently, if we subtract $p-1$ from $(p-1)!$, then the difference is divisible by p.

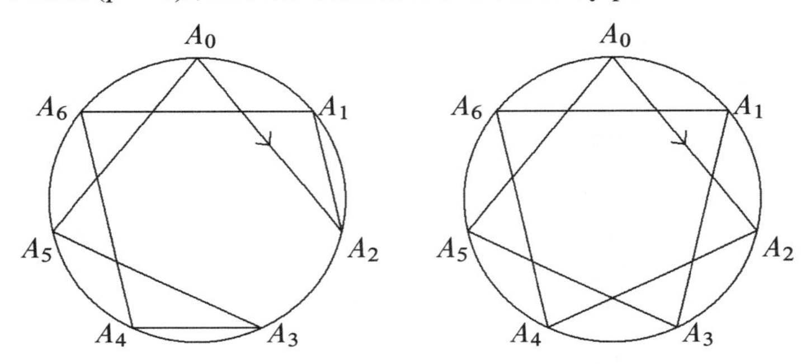

We classify the paths into groups as follows. If we obtain one path from another by a rotation about the center of the circle, we put them in the same group. Normally, each group consists of p paths. This is because A_0 can coincide with any of the p points, including itself. However, the diagram above on the right shows a path which is in a group all by itself.

When and how often does this happen? Suppose the path obtained is the same as the original one after a rotation of an angle $k\theta$, where $\theta = (\frac{360}{p})^\circ$ and k is a positive integer less than p. Then the path obtained will again be the same as the original one after a rotation of an angle $0, k\theta, 2k\theta, \ldots, (p-1)k\theta$.

We claim that no two of these p angles differ by a multiple of 360°. Suppose $0 \le i < j \le p-1$ and $jk\theta - ik\theta$ is a multiple of 360°. Then $\frac{k(j-i)}{p}$ is an integer, which is not possible since p is prime and both k and $j-i$ lie between 0 and p. This justifies the claim.

So we have the same path p times, which means that we have a path in a group all by itself. This happens when $k = 1, 2, \ldots, p-1$, for a total of $p-1$ times. These are the $p-1$ regular p-gons, two of which are convex polygons and the other $p-3$ star-polygons. This completes the proof of Wilson's Theorem.

As an application, we prove that if $p = 4k+1$ is a prime number, then there exists an integer n such that n^2+1 is divisible by p. By Wilson's Theorem, we have, in congruence modulo p,

$$\begin{aligned} -1 &\equiv (p-1)! \\ &= (4k)! \\ &\equiv 1 \times 2 \times \cdots \times 2k(-2k)(-(2k-1))\cdots(-1) \\ &= (-1)^{2k}((2k)!)^2 \\ &= ((2k)!)^2. \end{aligned}$$

It follows that p divides n^2+1 where $n = (2k)!$.

We now return to the problem of expressing positive integers as a sum of two squares. With the help of the result above, we can now prove that if $p = 4k+1$ is a prime number, then p can be expressed as the sum of two squares.

So let n be a positive integer such that n^2+1 is divisible by p. Let $t = \lfloor\sqrt{p}\rfloor$. Let each of x and y be one of 0, 1, 2, ..., t. Then the number of different expressions of the form $nx-y$ is $(t+1)^2 > (\sqrt{p})^2 = p$. Hence two of them must be congruent modulo p.

Let them be $nx_1 - y_1 \equiv nx_2 - y_2 \pmod{p}$. Then we have $n(x_1-x_2) \equiv y_1 - y_2 \pmod{p}$. If $x_1 = x_2$, then $y_1 - y_2 \equiv 0 \pmod{p}$. This is only possible if $y_1 = y_2$, but then the two expressions are not different. Hence $x_1 \neq x_2$, and we may assume by symmetry that $x_1 > x_2$.

Let $u = x_1 - x_2$ and $v = |y_1 - y_2|$. Then $0 < u < \sqrt{p}$ and $0 < v < \sqrt{p}$. We have $nu \equiv v \pmod{p}$. Since n^2+1 is divisible by p, $0 \equiv n^2+1 \equiv n^2u^2 + u^2 = v^2 + u^2 \pmod{p}$. Since $0 < v^2+u^2 < 2p$, we must have $p = v^2+u^2$, which is the desired result.

Once again, we digress and consider a companion result to Wilson's Theorem known as **Fermat's Little Theorem**. It states that for any prime number p and any positive integer n, $n^p - n$ is divisible by p. Moreover, if n is not divisible by p, then $n^{p-1} - 1$ is.

We now prove Fermat's Little Theorem. There are p points evenly spaced around a circle. We paint each of them in one of n colors, and count the number of distinct patterns up to rotation. Without rotation, there are n^p distinct designs since each of the p points can be painted with any of the n colors. There are n designs in which all points are painted in the same color, and these n designs yield n distinct patterns under rotation. In the remaining $n^p - n$ designs, not all the points have the same color. Since p is a prime number, each distinct pattern under rotation corresponds to p designs. Hence $n + \frac{n^p - n}{p}$ is the number of distinct patterns. It follows that $n^p - n$ must be divisible by p.

We now prove that if $p = 4k+3$ is a prime number, then it is not expressible as the sum of two squares. Suppose to the contrary that $p = u^2 + v^2$ for some positive integers u and v less than $\sqrt{p}$. Then $u^2 \equiv -v^2 \pmod{p}$, so that $1 \equiv u^{p-1} = (u^2)^{2k+1} \equiv (-v^2)^{2k+1} = -v^{p-1} \equiv -1 \pmod{p}$, which is a contradiction.

It follows that if the prime number $p = 4k + 3$ divides the sum of the squares of two integers, then p must divide each of these two integers so that the sum of two squares is divisible by p^2. Factoring out p^2, we still have a sum of two squares. It follows that a positive integer is expressible as the sum of two squares if and only if in its prime factorization, each prime number of the form $4k + 3$ appears an even number of times.

So not every positive integer can be expressed as the sum of two squares. It is not much more difficult to prove that positive integers of the form $4^t(8k + 7)$ cannot be expressed as the sum of three squares. The general problem, known as **Waring's Problem**, arose from his unproved statement in 1770 that every positive integers is expressible as the sum of at most 4 squares, as the sum of at most 9 cubes, as the sum of 19 fourth powers, and so on. What this means is that for any positive integer m, there exists a positive integer $g(m)$ such that every positive integer is expressible as the sum of at most $g(m)$ mth powers.

That $g(2) = 4$ was proved by Lagrange, a contemporary mathematician of Waring. The number 23 shows that $g(3) \geq 9$ and the number 79 shows that $g(4) \geq 19$. It was not until over a hundred years later that Hilbert gave the general solution, establishing the existence of the numbers $g(m)$. In his work, Hilbert obtained many spectacular results, and developed new methods that have found wide applications in the study of number theory.

A more significant function is $G(m)$, where every sufficiently large positive integer is expressible as the sum of at most $G(m)$ mth powers. This is essentially saying that the positive integers which cannot be expressed as the sum of less than $g(m)$ mth powers are finite in number, so that from a certain point on, less mth powers are sufficient. We always have $G(m) \leq g(m)$. It is known that $G(4) = 16$ while $19 \leq g(4) \leq 27$.

Finally, we look back at another problem from Hungarian Problem Book III.

Problem 1940.2

Let m and n be distinct positive integers. Prove that $2^{2^m}+1$ and $2^{2^n}+1$ are relatively prime to each other.

The numbers $F(k) = 2^{2^k}+1$ are called **Fermat Numbers**. The first five, $F(0) = 3$, $F(1) = 5$, $F(2) = 17$, $F(3) = 257$ and $F(4) = 65537$, are all prime numbers. Fermat boldly stated that all such numbers are prime numbers, but Euler disproved the conjecture by showing that $F(5) = 641 \times 6700417$.

A lot of work has been done on determining whether a specific Fermat number is a prime number of not. Many turn out to be composite, but not another prime number has been found among the Fermat numbers. It is not known whether there are finitely many or infinitely many Fermat primes.

The problem above states that every two Fermat numbers are relatively prime to each other. Whether they themselves are prime numbers or not, the prime numbers that appear in the prime factorization of one Fermat number cannot appear in the prime factorization of another Fermat number. Since there are infinitely many Fermat numbers, we can conclude that there are infinitely many prime numbers.

Here is a general construction of sequences of pairwise relatively prime numbers. Let λ and μ be relatively prime to each other. Choose $a_{-1} = 1$ and for $n \geq 1$, define $a_n = \lambda a_{n-1}a_{n-2}\cdots a_0 a_{-1} + \mu$. If we take $\lambda = 1$ and $\mu = 2$, we have the Fermat numbers.

Fermat numbers come up unexpectedly in the problem of constructing regular polygons by straight edge and compass. It turns out that a regular n-gon is constructible if and only if n is the product of a power of 2 and any number of distinct Fermat prime numbers.

4.3 Discussion on Algebra

We begin our discussion on Algebra by looking back at the following geometry problem from Hungarian Problem Book I.

Problem 1899.1

The points A_0, A_1, A_2, A_3, A_4 divide a unit circle into five equal parts. Prove that $(A_0A_1 \cdot A_0A_2)^2 = 5$.

After the Second Solution to this problem, we found "Note 2 — A geometric application of complex numbers". Certain properties of polynomials were used.

Factor Theorem Let $f(x) = a_0x^n + a_1x^{n-1} + \cdots + a_{n-1}x + a_n$ be a polynomial with real coefficients, where $a_0 \neq 0$ and $n \geq 1$. Suppose r is a real root of $f(x)$. Then $f(x) = (x - r)g(x)$ for some polynomial $g(x)$ of degree $n - 1$.

Proof Note that

$$f(r) = a_0r^n + a_1r^{n-1} + \cdots + a_{n-1}r + a_n,$$

so that

$$f(x) - f(r) = a_0(x^n - r^n) + a_1(x^{n-1} - r^{n-1}) + \cdots + a_{n-1}(x - r).$$

Note that

$$x^k - r^k = (x - r)(x^{k-1} + x^{k-2}r + \cdots + xr^{k-2} + r^{k-1}).$$

If r is a root of $f(x)$, then

$$f(x) = f(x) - f(r) = (x - r)g(x)$$

where

$$\begin{aligned} g(x) = a_0(&x^{n-1} + x^{n-2}r + \cdots + xr^{n-2} + r^{n-1}) \\ &+a_1(x^{n-2} + x^{n-3}r + \cdots + xr^{n-3} + r^{n-2}) \\ &+ \cdots + a_{n-1}. \end{aligned}$$

Since $a_0 \neq 0$, $g(x)$ is indeed of degree $n - 1$.

Suppose $r_1, r_2, \ldots, r_m$ are distinct real roots of $f(x)$. Applying the Factor Theorem repeatedly, we have

$$f(x) = (x - r_1)(x - r_2) \cdots (x - r_m)g(x)$$

for some polynomial $g(x)$ of degree $n - m$.

Clearly, we cannot have $m > n$. If $m = n$, then we have $g(x) = a_0$. If $m < n$, either $g(x)$ has no real roots, or real roots which are also roots of $f(x)$. In the latter case, we say that $f(x)$ has **multiple roots**, and the number of times a root appears is called its **multiplicity**. Suppose r_k has multiplicity s_k. Then

$$f(x) = (x - r_1)^{s_1}(x - r_2)^{s_2} \cdots (x - r_k)^{s_m}h(x)$$

for some polynomial $h(x)$ of degree $n - (s_1 + s_2 + \cdots + s_m) \geq 0$ which has no real roots.

Suppose we allow the coefficients of $f(x)$ to be complex numbers and consider its complex roots. Then Gauss proved that every polynomial of degree at least 1 has a root, from which it is easy to derive that every polynomial of degree $n \geq 1$ has exactly n roots, counting multiplicities. This important result is known as the **Fundamental Theorem of Algebra**.

Polynomials also turn up in the following problem from Hungarian Problem Book II.

Problem 1923.2

If

$$s_n = 1 + q + q^2 + \cdots + q^n$$

and

$$S_n = 1 + \frac{1+q}{2} + \left(\frac{1+q}{2}\right)^2 + \cdots + \left(\frac{1+q}{2}\right)^n,$$

prove that

$$\binom{n+1}{1} + \binom{n+2}{2}s_1 + \cdots + \binom{n+1}{n+1}s_n = 2^n S_n.$$

In the Second Solution, the cases $q = 1$ and $q \neq 1$ were treated separately. Actually, the special case $q = 1$ can be derived from the general case $q \neq 1$ by appealing to the following result.

Identical Polynomials Theorem If two polynomials of degree at most n agree on more than n values, then they are the same polynomial.

Proof Let the two polynomials be $f(x)$ and $g(x)$. Each value on which they agree is a root of the polynomial $f(x)-g(x)$. Hence it has more than n roots. Since each of $f(x)$ and $g(x)$ has degree at most n, so has $f(x)-g(x)$. This is only possible if $f(x)-g(x)$ is identically zero. In other words, $f(x)$ and $g(x)$ are the same polynomial.

Let $f(x) = a_0x^n + a_1x^{n-1} + \cdots + a_{n-1}x + a_n$ and let $r_1, r_2, \ldots, r_n$ be its roots. We have already proved that we have

$$f(x) = a_0(x - r_1)(x - r_2)\cdots(x - r_n).$$

Expanding and equating coefficients of corresponding terms, we have

$$\begin{aligned} r_1 + r_2 + \cdots + r_n &= -\frac{a_1}{a_0}, \\ r_1r_2 + r_1r_3 + \cdots + r_{n-1}r_n &= \frac{a_2}{a_0}, \\ r_1r_2r_3 + r_1r_2r_4 + \cdots + r_{n-2}r_{n-1}r_n &= -\frac{a_3}{a_0}, \\ \cdots &= \cdots, \\ r_1r_2\cdots r_n &= (-1)^n\frac{a_n}{a_0}. \end{aligned}$$

These are known collectively as **Vieta's Formulae**. The left side of the kth formula is the sum of all products of the roots taken k at a time. The right side of the kth formula is $(-1)^k\frac{a_k}{a_0}$.

We now turn to the following problem from Hungarian Problem Book III.

Problem 1935.1

Let n be a positive integer. Prove that

$$\frac{a_1}{b_1} + \frac{a_2}{b_2} + \cdots + \frac{a_n}{b_n} \geq n,$$

where $\langle b_1, b_2, \ldots, b_n\rangle$ is any permutation of the positive real numbers a_1, $a_2, \ldots, a_n$.

The simplest solution of this problem makes use of the **Rearrangement Inequality**. Let $a_1 \leq a_2 \leq \cdots \leq a_n$ and $b_1 \leq b_2 \leq \cdots \leq b_n$ be real numbers. Suppose $\langle c_1, c_2, \ldots, c_n\rangle$ is any permutation of $b_1, b_2, \ldots, b_n$. Then

$$\begin{aligned} &a_1b_n + a_2b_{n-1} + \cdots + a_nb_1 \\ &\leq a_1c_1 + a_2c_2 + \cdots + a_nc_n \\ &\leq a_1b_1 + a_2b_2 + \cdots + a_nb_n. \end{aligned}$$

Here is a very simple application. Suppose there are \$10 bills in the first box, \$20 dollar bills in the second box, \$50 bills in the third box and \$100 bills in the fourth box. You are allowed to take 3, 4, 5 and 6 bills from different boxes. How should you maximize the amount you take? Obviously, you should take 6 bills from the fourth box, 5 bills from the third box, 4 bills from the second box and 3 bills from the first box, for a total of \$960.

We had pointed out the fundamental importance of this result, and had used it to prove Cauchy's Inequality. We now give further demonstration of its power by deriving several other classical inequalities.

The Arithmetic-Geometric Means Inequality *For any positive real numbers $x_1, x_2, \ldots, x_n$, we have*

$$\frac{x_1 + x_2 + \cdots + x_n}{n} \geq \sqrt[n]{x_1 x_2 \cdots x_n}.$$

Equality holds if and only if $x_1 = x_2 = \cdots = x_n$.

Proof Let $G = \sqrt[n]{x_1 x_2 \cdots x_n}$. Let

$$a_1 = \frac{x_1}{G},\ a_2 = \frac{x_1 x_2}{G^2}, \ldots \quad \text{and} \quad a_n = \frac{x_1 x_2 \cdots x_n}{G^n} = 1.$$

Let $b_k = \frac{1}{a_k}$ for $1 \leq k \leq n$. Since $b_i < b_j$ if and only if $a_i > a_j$, the Rearrangement Inequality yields

$$a_1 b_1 + a_2 b_2 + \cdots + a_n b_n \leq a_1 b_n + a_2 b_1 + \cdots + a_n b_{n-1}.$$

In other words,

$$n = 1 + 1 + \cdots + 1 \leq \frac{x_1}{G} + \frac{x_2}{G} + \cdots + \frac{x_n}{G}.$$

It follows that

$$G \leq \frac{x_1 + x_2 + \cdots + x_n}{n}.$$

Equality holds if and only if $b_n = b_1 = b_2 = \cdots = b_{n-1}$. Hence

$$\frac{x_1}{G} = \frac{x_2}{G} = \cdots = \frac{x_n}{G},$$

which is equivalent to $x_1 = x_2 = \cdots = x_n$.

Chebyshev's Inequality Let $a_1 \leq a_2 \leq \cdots \leq a_n$ be real numbers. For real numbers $b_1 \leq b_2 \leq \cdots \leq b_n$, we have

$$\frac{a_1 b_1 + a_2 b_2 + \cdots + a_n b_n}{n} \geq \frac{a_1 + a_2 + \cdots + a_n}{n} \cdot \frac{b_1 + b_2 + \cdots + b_n}{n}.$$

For real numbers $b_1 \geq b_2 \geq \cdots \geq b_n$, we have

$$\frac{a_1 b_1 + a_2 b_2 + \cdots + a_n b_n}{n} \leq \frac{a_1 + a_2 + \cdots + a_n}{n} \cdot \frac{b_1 + b_2 + \cdots + b_n}{n}.$$

Proof By the Rearrangement Inequality, we have

$$\begin{aligned}
a_1 b_1 + a_2 b_2 + \cdots + a_n b_n &= a_1 b_1 + a_2 b_2 + \cdots + a_n b_n, \\
a_1 b_1 + a_2 b_2 + \cdots + a_n b_n &\geq a_1 b_2 + a_2 b_3 + \cdots + a_n b_1, \\
a_1 b_1 + a_2 b_2 + \cdots + a_n b_n &\geq a_1 b_3 + a_2 b_4 + \cdots + a_n b_2, \\
\cdots &\geq \cdots, \\
a_1 b_1 + a_2 b_2 + \cdots + a_n b_n &\geq a_1 b_n + a_2 b_1 + \cdots + a_n b_{n-1}.
\end{aligned}$$

Addition yields

$$n(a_1b_1 + a_2b_2 + \cdots + a_nb_n) \geq (a_1 + a_2 + \cdots + a_n)(b_1 + b_2 + \cdots + b_n).$$

This is equivalent to the first desired inequality. The second desired inequality can be proved in an analogous manner.

We point out that the Root-Mean Square Arithmetic Mean Inequality, which we had introduced back in Chapter II, is a simple corollary of Chebyshev's Inequality. Another result we introduced there is Jensen's Inequality. We now generalize this to Jensen's Inequality for two variables.

A function $f(x, y)$ is said to be **convex(concave)** over a rectangle if a chord joining any two points on the graph of the function within the rectangle lies on or above (below) the graph itself. For most functions, it is sufficient to know that the midpoint of the chord lies on or above(below) the graph. This may be expressed algebraically as

$$f\left(\frac{x_1 + x_2}{2}, \frac{y_1 + y_2}{2}\right) \leq (\geq) \frac{f(x_1, y_1) + f(x_2, y_2)}{2}.$$

Jensen's Inequality then states that for a concave function $f(x)$ over an interval,

$$f(\frac{x_1 + x_2 + \cdots + x_n}{n}, \frac{y_1 + y_2 + \cdots + y_n}{2}) \leq (\geq) \frac{f(x_1, y_1) + f(x_2, y_2) + \cdots + f(x_n, y_n)}{n},$$

where $(x_1, y_1), (x_2, y_2), \ldots, (x_n, y_n)$ are points in this rectangle. The proof is analogous to that in Chapter II.

The second solution to the following problem from Hungarian Problem Book III essentially proves that the function $\sqrt{xy}$ is concave, furnishing us a first example on functions of two variables.

Problem 1939.1

Let a_1, a_2, b_1, b_2, c_1 and c_2 be real numbers for which $a_1a_2 > 0$, $a_1c_1 \geq b_1^2$ and $a_2c_2 \geq b_2^2$. Prove that

$$(a_1 + a_2)(c_1 + c_2) \geq (b_1 + b_2)^2.$$

Using this, we can offer another proof of the following result.

Cauchy's Inequality Let $a_1, a_2, \ldots, a_n, b_1, b_2, \ldots, b_n$ be real numbers. Then

$$(a_1b_1 + a_2b_2 + \cdots + a_nb_n)^2 \leq (a_1^2 + a_2^2 + \cdots + a_n^2)(b_1^2 + b_2^2 + \cdots + b_n^2),$$

with equality if and only if for some constant k, $a_i = kb_i$ for $1 \leq i \leq n$ or $b_i = ka_i$ for $1 \leq i \leq n$.

Proof Let $x_i = a_i^2$ and $y_i = b_i^2$ for $1 \le i \le n$. By Jensen's Inequality,

$$\begin{aligned}
&\frac{\sqrt{x_1y_1} + \sqrt{x_2y_2} + \cdots + \sqrt{x_ny_n}}{n} \\
&\quad \le \sqrt{\frac{x_1 + x_2 + x \cdots + x_n}{n} \cdot \frac{y_1 + y_1 + \cdots + y_n}{n}} \\
&\quad = \frac{\sqrt{(x_1 + x_2 + \cdots + x_n)(y_1 + y_2 + \cdots + y_n)}}{n}.
\end{aligned}$$

This is equivalent to the desired result. We omit the checking of the condition on equality.

The following problem from Hungarian Problem Book I provides a nice transition from our current topic on inequalities to the next topic, on infinity. The two topics are intimately related since approaching infinity means growing bigger than any fixed value.

Problem 1905.2

Divide the unit square into 9 equal squares by means of two pairs of lines parallel to the sides. Now remove the central square. Treat the remaining 8 squares the same way, and repeat this process n times.

(a) How many squares of side length $\frac{1}{3^n}$ remain?

(b) What is the sum of the areas of the removed squares as n becomes infinite?

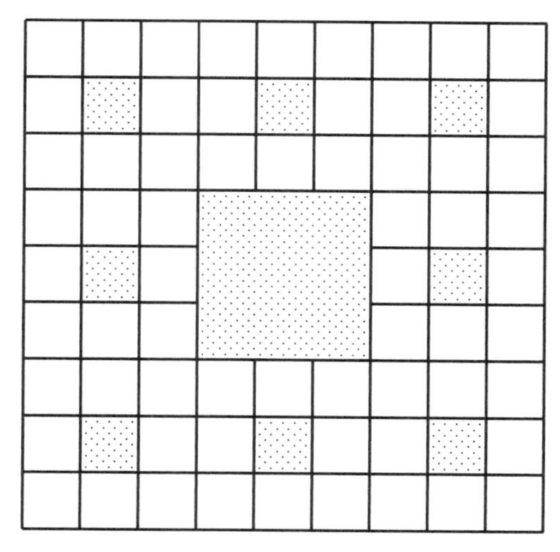

In the solution to part (b), we need to show that as n approaches infinity, $(\frac{8}{9})^n$ approaches 0. In other words, $(\frac{9}{8})^n$ approaches infinity. This follows from **Bernoulli's Inequality** which states that $(1+x)^n \geq 1+nx$ for all $x \geq -1$ and all $n \geq 1$. Then $(\frac{9}{8})^n = (1+\frac{1}{8})^n \geq 1+\frac{n}{8}$, and the last term clearly approaches infinity as n does. Bernoulli's Inequality can be produced by mathematical induction on n.

Next, we turn our attention to the following problem in Hungarian Problem Book III.

Problem 1936.1

Prove that for all nonnegative integers n,

$$\frac{1}{1\cdot 2}+\frac{1}{3\cdot 4}+\cdots+\frac{1}{(2n-1)2n}=\frac{1}{n+1}+\frac{1}{n+2}+\cdots+\frac{1}{2n}.$$

Suppose we let the expression on the left side continue indefinitely beyond $\frac{1}{(2n-1)2n}$, so that we have

$$\frac{1}{1\cdot 2}+\frac{1}{3\cdot 4}+\cdots+\frac{1}{(2n-1)2n}+\frac{1}{(2n+1)(2n+2)}+\cdots.$$

Such an expression is called an **infinite series**. Here are some more.

(1) $1+\frac{1}{2}+\frac{1}{4}+\frac{1}{8}+\cdots+\frac{1}{2^{n-1}}+\cdots,$

(2) $2+\frac{3}{2}+\frac{9}{8}+\frac{27}{32}+\cdots+2(\frac{3}{4})^{n-1}+\cdots,$

(3) $1-0.9+0.81-0.729+\cdots+(-0.9)^{n-1}+\cdots,$

(4) $0.23+0.0023+0.000023+\cdots,$

(5) $\frac{1}{1\cdot 2}+\frac{1}{2\cdot 3}+\frac{1}{3\cdot 4}+\frac{1}{4\cdot 5}+\frac{1}{5\cdot 6}+\cdots,$

(6) $2+\frac{3}{2}+\frac{4}{3}+\frac{5}{4}+\frac{6}{5}+\frac{7}{6}+\cdots.$

At the moment, an infinite series has no meaning because we have no way of adding infinitely many numbers. However, it is worth our while to pursue this matter, in order to come up with a sensible definition of the sum of an infinite series. This issue has its root in arithmetic. Example (4) is simply an unusual way of writing the recurrent decimal 0.232323.... When this number is involved in computations, we usually truncate it at some point. Exactly where the truncation occurs depends on the degree of accuracy required. The main point is that when necessary, we can truncate it and obtain an approximate value, such that the error can be made as small as desired.

Let us try to find something that may be considered as the sum of the infinite series above. The sum of the first n terms is called the nth **partial sum**, which can meaningfully be computed. Examples (1) to (4) are examples of **geometric series**, where each term after the first is obtained from the preceding one by multiplying a fixed number called the **common ratio**. The common ratio in (1) is $\frac{1}{2}$, that in (2) is $\frac{3}{4}$, that in (3) is -0.9 and that in (4) is 0.01. Let us calculate the nth partial sum S of (1).

$$S = 1 + \frac{1}{2} + \frac{1}{4} + \frac{1}{8} + \cdots + \frac{1}{2^{n-1}},$$
$$\frac{1}{2}S = \frac{1}{2} + \frac{1}{4} + \frac{1}{8} + \cdots + \frac{1}{2^{n-1}} + \frac{1}{2^n}.$$

Subtraction yields

$$\frac{1}{2}S = 1 - \frac{1}{2^n}$$

so that the nth partial sum is $2-2(\frac{1}{2})^n$. In an analogous manner, we find that the nth partial sums in (2), (3) and (4) are

$$8 - 8(\frac{3}{4})^n, \quad \frac{10}{19} - \frac{10}{19}(-\frac{9}{10})^n \quad \text{and} \quad \frac{23}{99} - \frac{23}{99}(\frac{1}{100})^n,$$

respectively.

Each of these expressions contains a number with absolute value less than 1, which is raised to the nth power. As n increases, this becomes arbitrarily small, and this is unchanged by its multiplication with a fixed number such as 2, 8, $\frac{10}{19}$ and $\frac{23}{99}$. We say that such an infinite series is **convergent**, and it is meaningful to assign it a value. They are 2, 8, $\frac{10}{19}$ and $\frac{23}{99}$ for (1), (2), (3) and (4) respectively.

In general, the infinite series $a_1 + a_2 + \cdots + a_n + \cdots$ is convergent to a real number S if for any real number $\epsilon > 0$, there exists an N such that whenever $n > N$, the absolute value of the difference between S and the nth partial sum of the infinite series is less than ϵ. In such a case, we often write $a_1 + a_2 + \cdots + a_n + \cdots = S$, with the understanding that the infinite sum is interpreted according to our definition.

As we have seen, the infinite series maintains most of the properties of the partial sums, but not all. Not every infinite series has a finite sum. For instance, in (6), each term is greater than 1. When n is sufficiently large, we can make the nth partial sum larger than any prescribed number. It follows that there is no real number which satisfies the condition in the definition above, and the series is not convergent. We said that such an infinite series is **divergent**.

For an infinite series to be divergent, it is not necessary for the partial sums, or even the absolute values of the partial sums, to increase without bound. For instance, the infinite series $1-1+1-1+1-1+\cdots$ is divergent, since the even partial sums have value 0 while the odd partial sums have value 1. There is no value that can be arbitrarily close to both 0 and 1.

We now return to infinite series (5). At first glance, it is not easy to determine its nth partial sum. However, if we notice that the kth term $\frac{1}{k(k+1)}$ may be expressed as $\frac{1}{k}-\frac{1}{k+1}$, then the nth partial sum becomes

$$\begin{aligned}
&\frac{1}{1\times 2}+\frac{1}{2\times 3}+\frac{1}{3\times 4}+\cdots+\frac{1}{(n-1)n}+\frac{1}{n(n+1)}\\
&\quad=\left(1-\frac{1}{2}\right)+\left(\frac{1}{2}-\frac{1}{3}\right)+\left(\frac{1}{3}-\frac{1}{4}\right)\\
&\qquad+\cdots+\left(\frac{1}{n-1}-\frac{1}{n}\right)+\left(\frac{1}{n}-\frac{1}{n+1}\right)\\
&\quad=1-\frac{1}{n+1}.
\end{aligned}$$

When n is sufficiently large, $\frac{1}{n+1}$ is arbitrarily small. It follows that

$$\frac{1}{1\times 2}+\frac{1}{2\times 3}+\frac{1}{3\times 4}+\frac{1}{4\times 5}+\cdots=1.$$

We have only examined a few examples of infinite series whose convergence can be determined relatively easily. The general problem is much more difficult. In the second solution to Problem 1936.1, the left side of the desired result has been expressed as

$$\begin{aligned}
&\left(1+\frac{1}{2}+\cdots+\frac{1}{2n}\right)-\left(1+\frac{1}{2}+\cdots+\frac{1}{n}\right)\\
&\quad=\left(1+\frac{1}{2}+\cdots+\frac{1}{2n}\right)-2\left(\frac{1}{2}+\frac{1}{4}+\cdots+\frac{1}{2n}\right)\\
&\quad=1-\frac{1}{2}+\frac{1}{3}-\frac{1}{4}+\cdots+\frac{1}{2n-1}-\frac{1}{2n}.
\end{aligned}$$

This is none other than the $2n$th partial sum of the infinite series $1-\frac{1}{2}+\frac{1}{3}-\frac{1}{4}+\cdots+\frac{1}{2n-1}-\frac{1}{2n}+\cdots$. We shall now prove that this infinite series is convergent.

Extracting a common factor $\frac{1}{n}$ from the right side of the desired result in Problem 1936.1, we have

$$\frac{1}{n}\left(\frac{1}{1+\frac{1}{n}}+\frac{1}{1+\frac{2}{n}}+\cdots+\frac{1}{1+\frac{n}{n}}\right).$$

This can be interpreted graphically as follows

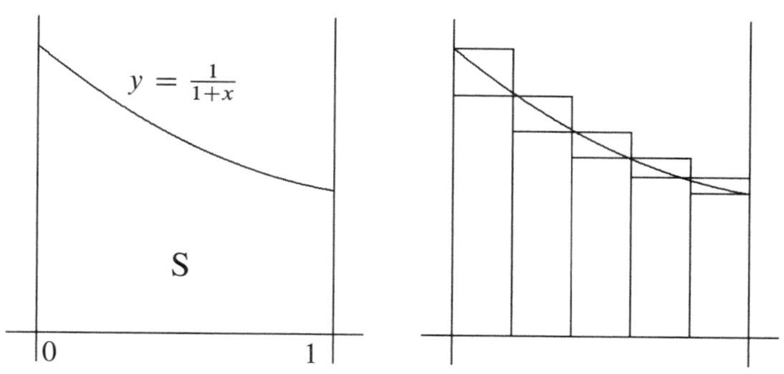

Consider the function $y = \frac{1}{1+x}$ on the interval [0,1]. Let S denote the area under the curve. We estimate S as follows. Divide the interval into n equal parts. Connect each division point by a vertical line to the curve, so that the area is divided into n trapezoids with curved tops, each having width $\frac{1}{n}$. Approximate each trapezoid by two rectangles. The sum of the areas of the larger rectangles is

$$\frac{1}{n} + \frac{1}{n+1} + \cdots + \frac{1}{2n-1}$$

while the sum of the areas of the smaller rectangles is

$$\frac{1}{n+1} + \frac{1}{n+2} + \cdots + \frac{1}{2n}.$$

The value of S lies between these two sums, which differ by $\frac{1}{n}$. It follows that when n is sufficiently large, the difference between S and the $2n$th partial sum of the infinite series $1 - \frac{1}{2} + \frac{1}{3} - \frac{1}{4} + \cdots + \frac{1}{2n-1} - \frac{1}{2n} + \cdots$ is arbitrarily small. Those familiar with Calculus will know that

$$S = \int_0^1 \frac{1}{1+x} dx = \ln 2.$$

Suppose we rearrange the terms of the above infinite series by taking groups of one positive term followed by two negative terms. Thus we have

$$\begin{aligned} S &= \left(1 - \frac{1}{2} - \frac{1}{4}\right) + \left(\frac{1}{3} - \frac{1}{6} - \frac{1}{8}\right) + \left(\frac{1}{5} - \frac{1}{10} - \frac{1}{12}\right) \\ &\quad + \cdots + \left(\frac{1}{2n-1} - \frac{1}{4n-2} - \frac{1}{4n}\right) \\ &= \frac{1}{2} - \frac{1}{4} + \frac{1}{6} - \frac{1}{8} + \frac{1}{10} - \frac{1}{12} + \cdots + \frac{1}{4n-2} - \frac{1}{4n} \\ &= \frac{1}{2} S. \end{aligned}$$

This is a contradiction since $S \neq 0$.

The reason we get into this trouble is that the infinite series $1-\frac{1}{2}+\frac{1}{3}-\frac{1}{4}+\cdots$ is only **conditionally convergent**, in that the corresponding infinite series with absolute values $1+\frac{1}{2}+\frac{1}{3}+\frac{1}{4}+\cdots$ is divergent. If the corresponding infinite series with absolute values is convergent, then the original series is said to be **absolutely convergent**, and a rearrangement of the terms of an absolutely convergent infinite series will not affect its value.

The infinite series $1+\frac{1}{2}+\frac{1}{3}+\frac{1}{4}+\cdots$ is a famous one called the **harmonic series**. We can show that its partial sums increase without bound. Let S_k denote its kth partial sum. Then

$$\begin{aligned} S_{2k}-S_k &= \frac{1}{k+1}+\frac{1}{k+2}+\cdots+\frac{1}{2k} \\ &> \frac{1}{2k}+\frac{1}{2k}+\cdots+\frac{1}{2k} \\ &= \frac{1}{2}. \end{aligned}$$

In other words,

$$\begin{aligned} &1+\frac{1}{2}+\frac{1}{3}+\frac{1}{4}+\cdots \\ &=1+\frac{1}{2}+\left(\frac{1}{3}+\frac{1}{4}\right)+\left(\frac{1}{5}+\frac{1}{6}+\frac{1}{7}+\frac{1}{8}\right)+\cdots \\ &>1+\frac{1}{2}+\frac{1}{2}+\frac{1}{2}+\cdots. \end{aligned}$$

It follows that the harmonic series is divergent.

Note that $S_{4k}-S_k=(S_{4k}-S_{2k})+(S_{2k}-S_k)>\frac{1}{2}+\frac{1}{2}=1$. Hence

$$\frac{1}{k+1}+\frac{1}{k+2}+\cdots+\frac{1}{4k}>1,$$

so that whenever $m \geq 4k$,

$$\frac{1}{k+1}+\frac{1}{k+2}+\cdots+\frac{1}{m}>1.$$

This gives us an alternative solution to the following problem in Hungarian Problem Book III.

Problem 1938.2

Prove that for all integers $n>1$,

$$\frac{1}{n}+\frac{1}{n+1}+\cdots+\frac{1}{n^2-1}+\frac{1}{n^2}>1.$$

For $k = n-1$ and $m = n^2$, $m \geq 4k$ is equivalent to $n^2 \geq 4(n-1)$, which is true since $(n-2)^2 \geq 0$. The desired result follows immediately.

Let T_n denote the sum in Problem 1938.2. Then

$$\begin{aligned} T_n &> \left(\frac{1}{n+1} + \cdots + \frac{1}{2n}\right) + \left(\frac{1}{2n+1} + \cdots + \frac{1}{3n}\right) \\ &\quad + \cdots + \left(\frac{1}{(n-1)n+1} + \cdots + \frac{1}{n^2}\right) \\ &> n\left(\frac{1}{2n} + \frac{1}{3n} + \cdots + \frac{1}{n^2}\right) \\ &= S_n - 1. \end{aligned}$$

Since S_n increases without bound as n increases, so does T_n.

We now turn to a different aspect of infinity. This motivated by the following problem in Hungarian Problem Book III.

Problem 1936.3

Let a be any positive integer. Prove that there exists a unique pair of positive integers x and y such that

$$x + \frac{1}{2}(x+y-1)(x+y-2) = a.$$

In the second solution to this problem, it was established that the set of ordered pairs of positive integers can be listed in a sequence as follows:

$$(1,1), (1,2), (2,1), (1,3), (2,2), (3,1), \ldots,$$

In other words, it can be put into a one-to-one correspondence with the set of positive integers. This idea gives us a way of comparing sizes of infinite sets.

Suppose A and B are finite sets such that each element of A corresponds to a different element of B. If for each element of B, there is an element of A which corresponds to it, then we say that A and B are of the same size, or more formally, that they have the same **cardinality**. We write $|A| = |B|$. On the other hand, if there is at least one element of B to which no element in A corresponds, then we say that A is of a smaller size than B, or has a lower cardinality than B. We write $|A| < |B|$.

Now suppose A and B are infinite sets such that each element of A corresponds to a different element of B. If for each element of B, there is an element of A which corresponds to it, then we say that A and B have the same cardinality. However, if there is at least one element of B to which no element in A corresponds, it is not necessarily true that A has a lower cardinality than B.

Clearly, the set of positive integers can be put into a one-to-one correspondence with a subset of the ordered pairs of positive integers, namely, make n correspond to $(n, 1)$. However, the one-to-one correspondence exhibited in the second solution to Problem 1936.3 shows that the two sets have the same cardinality. Sets that have the same cardinality as the set of positive integers are said to be **countable**, because its elements can be listed in a sequence and be counted in order.

The set of positive rational numbers is countable. We consider each ordered pair (m, n) of positive integers as the fraction $\frac{m}{n}$, and remove those in which m and n have common factors greater than 1. Using the same one-to-one correspondence as before, we can list the positive rational numbers in a sequence as follows:

$$1, \frac{1}{2}, 2, \frac{1}{3}, 3, \frac{1}{4}, \frac{2}{3}, \frac{3}{2}, 4, \frac{1}{5}, 5, \ldots.$$

This sequence may be modified as follows to show that the set of all rational numbers is also countable:

$$0, 1, -1, \frac{1}{2}, -\frac{1}{2}, 2, -2, \frac{1}{3}, -\frac{1}{3}, 3, -3, \ldots.$$

The one-to-one correspondence between the set of positive integers and the set of ordered pairs of positive integers can be used to prove that if we have a sequence of countable sets $S_1, S_2, S_3, \ldots$, then their union is also countable. We simply make the nth element in S_m correspond to the ordered pair (m, n). In the case where the S_m are not pairwise disjoint, we just delete the common elements.

From this, we can deduce that the set of polynomials with integer coefficients 1 is countable. We first determine all polynomials such that the sum of the absolute values of their coefficients plus their degrees is a fixed value m. For instance, for $m = 4$, the only constant polynomials are 4 and -4, the only linear polynomials are $x + 2, x - 2, -x + 2, -x - 2$, $2x+1, 2x-1, -2x+1, -2x-1, 3x$ and $-3x$. The only quadratic polynomials are $x^2+1, x^2-1, -x^2+1, -x^2-1, x^2+x, x^2-x, -x^2+x, -x^2-x, 2x^2$ and $-2x^2$. The only cubic polynomials are x^3 and $-x^3$. There are no quartic polynomials since all coefficients must be zero.

Using the same method, we can list all the polynomials corresponding to any particular value of m. Starting with $m = 0, 1, 2, 3, 4, \ldots,$, we can list all such polynomials as follows:

$$0, 1, -1, 2, -2, x, -x, 3, -3, x + 1, x - 1, -x + 1, -x - 1,$$

$$2x, -2x, x^2, -x^2, 4, -4, x+2, x-2, -x+2, -x-2,$$
$$2x+1, 2x-1, -2x+1, -2x-1, 3x, -3x,$$
$$x^2+1, x^2-1, -x^2+1, -x^2-1, x^2+x, x^2-x,$$
$$-x^2+x, -x^2-x, 2x^2, -2x^2, x^3, -x^3, \ldots.$$

The roots of polynomials with integral coefficients are called **algebraic** numbers. Since the number of roots of a polynomial does not exceed its degree, they can be listed in place of the polynomials themselves. It follows that the set of algebraic numbers is countable.

Examples of countable infinite sets bring up a natural question. Are all infinite sets alike, or are there **uncountable** infinite sets? We answer this question by showing that the set S of real numbers in the interval (0,1) is uncountable.

We express each number in S as a decimal in the non-terminating form. This is because a terminating decimal such as 0.7352 can also be represented by 0.73529999..., since both are equal to $\frac{919}{1250}$. With this convention, every number in S has a unique expression as a non-terminating decimal.

The desired result can be established by showing that in any infinite sequence of numbers in S, at least one number in S will be missing.

So consider any infinite sequence of numbers in S, expressed as non-terminating decimals. Let the jth decimal digit of the ith number be denoted by $a_{i,j}$. Thus the sequence starts as follows:

$$0.a_{1,1}a_{1,2}a_{1,3}\ldots,$$
$$0.a_{2,1}a_{2,2}a_{2,3}\ldots,$$
$$0.a_{3,1}a_{3,2}a_{3,3}\ldots,$$
$$\ldots.$$

We now construct a number b in S which cannot belong to this sequence. Let its kth decimal digit be b_k, chosen so that it is not equal to $a_{k,k}$. More specifically, if $a_{k,k} \neq 5$, then we define $b_k = 5$, but if $a_{k,k} = 5$, then we define $b_k = 6$. Clearly, b differs from the kth number in the sequence in the kth decimal digit. It follows that b does not belong to this sequence. This argument is known as **Cantor's Diagonalization Method**.

There is nothing special about the interval (0, 1). We could have considered any interval (a, b) where $a < b$. In fact, we will show that the two intervals contain the same number of real numbers by exhibiting a one-to-one correspondence.

Draw a line segment AB representing the real numbers in the interval (a, b). Draw a parallel line segment CD representing the real numbers in the interval (0, 1). Let O be the point of intersection of AC and BD.

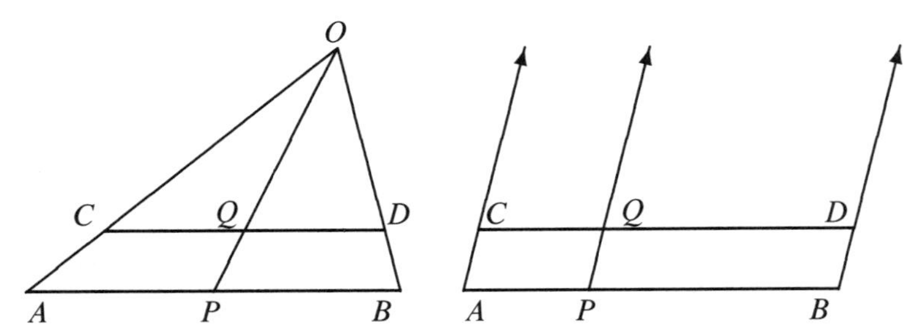

If a line through O intersects AB at some point P, it will also intersection CD at some point Q. Both points are uniquely determined by the line through O, and they correspond to each other. In the event that AC is parallel to BD, draw lines parallel to AC.

The number of real numbers in the interval (0,1) is actually equal to the number of all positive real numbers. Let O be the point with coordinates $(-1, 1)$. For any positive real number x, connect the point P with coordinates $(x, 0)$ to O, intersecting the interval (0,1) on the y-axis at the point Q with coordinates $(0, y)$, where $y = \frac{x}{1+x}$. This one-to-one correspondence establishes the desired result.

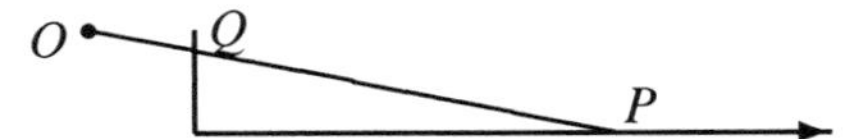

We can in fact show that the number of real numbers in any interval (a, b) with $a < b$ is equal to the number of all real numbers. We shift the interval (a, b) to the interval $(-\frac{b-a}{2}, \frac{b-a}{2})$ and place it on the y-axis. Let O be the point with coordinates $(-1, \frac{b-a}{2})$ and Z be the point with coordinates $(1, -\frac{b-a}{2})$. The real number 0 corresponds to 0 on the interval. A positive real number x corresponds $y = \frac{x}{1+x}$ on the interval as before. For any negative real number x, connect the point P with coordinates $(x, 0)$ to Z, intersecting the interval $(-\frac{b-a}{2}, 0)$ on the y-axis at the point Q with coordinates $(0, y)$, where $y = \frac{x}{1+x}$. This one-to-one correspondence establishes the desired result.

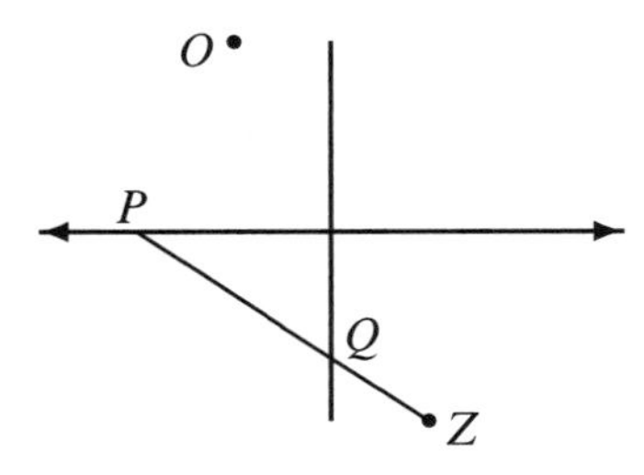

The geometric approach is somewhat problematic if we consider the intervals [0,1], [0,1) or (0,1] instead of (0,1). We will modify the argument as follows. From the formula $y = \frac{x}{1+x}$, we can see that x is rational if and only if y is rational. The correspondence between the irrational numbers can be handled in exactly the same way as before. As for the rational numbers, they are countable, and we can easily find an alternative correspondence between them.

The number of real numbers is referred to as the **continuum**. The number of elements of a countable set is less than the continuum because the elements of a countable set can be put into one-to-one correspondence with a subset of the real numbers (such as the integers or the rational numbers), but the set of real numbers is uncountable.

In particular, since we have earlier proved that the set of rational numbers is countable, we can conclude that the set of irrational numbers is uncountable. Similarly, since the set of algebraic numbers is countable, the set of non-algebraic or **transcendental** numbers is also uncountable.

It can be proved that there exist sets whose numbers of elements are greater than the continuum. In fact, the size of infinite sets increases without bound.

There is one question which arises naturally. Does there exist an infinite set which is uncountable, but whose number of elements is less than the continuum? Cantor's conjecture is that the answer is no, and this is known as the **Continuum Hypothesis**. However, it cannot be settled as it is possible to develop a consistent theory of sets in which the hypothesis is true, and also possible to develop an equally consistent theory of sets in which the hypothesis is false, much like the case with the Parallel Postulate in geometry.

4.4 Discussion on Geometry

Despite their appearances, the following two problems from the Hungarian Problem Book III are closely related, as was shown in the Second Solution to Problem 1942.2.

Problem 1941.2

Prove that if all four vertices of a parallelogram are lattice points and there are some other lattice points in or on the parallelogram, then its area exceeds 1.

Problem 1942.2

Let a, b, c and d be integers such that for all integers m and n, there exist integers x and y such that $ax + by = m$ and $cx + dy = n$. Prove that $ad - bc = \pm 1$.

As a reminder, a point whose coordinates are both integers is called a lattice point, and a polygon whose vertices are all lattice points is called a lattice polygon. A lattice polygon containing no lattice points other than its vertices is called a **basic** lattice polygon.

Let us prove directly that the area of a basic parallelogram is 1. If both pairs of opposite sides of the parallelogram are parallel to the coordinate axes, then the parallelogram is a unit square and its area is clearly 1. Suppose the opposite sides AD and BC of the parallelogram $ABCD$ are parallel to the y-axis. Then $AD = 1$ since there are no other lattice points between A and D. If the distance between AD and BC exceeds 1, then $ABCD$ contains between AD and BC a segment of length 1 of a vertical grid line. Either both end points of this segment are lattice points, or there is a lattice point in its interior. In either case, $ABCD$ cannot be a basic lattice parallelogram. It follows that the distance between AD and BC is exactly 1, and the area of $ABCD$ is 1.

Suppose no pair of opposite sides of the parallelogram $ABCD$ are parallel to the coordinate axes. Let AC be the longer diagonal. Let E be the point symmetric to C with respect to D. Then $ABDE$ is also a parallelogram with the same area as $ABCD$. Since $ABCD$ is a basic parallelogram, ABD is a basic triangle. Hence $ABDE$ is also a basic parallelogram.

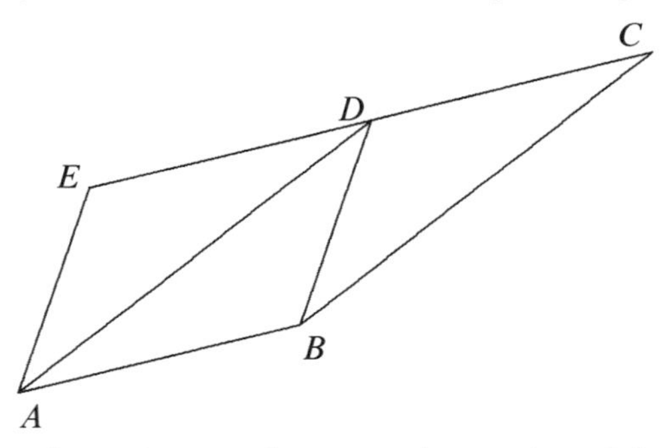

This transformation may be repeated a number of times, in two possible directions. Eventually, we obtain a basic parallelogram with one pair of opposite sides parallel to a coordinate axis, and we have proved that in this case, the area of the parallelogram is 1.

From this, it follows that the area of a basic lattice triangle is $\frac{1}{2}$. We now give an alternative proof, using the fact that the area of any lattice triangle is at least $\frac{1}{2}$.

For any basic lattice triangle, enclose it in a lattice rectangle and divide the remaining part of the rectangle into basic lattice triangles. We claim that the number of basic lattice triangle so obtained is constant. Let the vertices of the lattice rectangle be (0,0), $(m,0)$, (m,n) and $(0,n)$. We shall determine the sum of all interior angles of the basic lattice triangles.

The sum of the interior angles of the basic lattice triangles at each of these points is $\frac{\pi}{2}$. There are $2(m-1)+2(n-1)$ lattice points on the perimeter of the rectangle apart from the four vertices. The sum of the interior angles of the basic lattice triangles at each of these points is π because such a lattice point cannot be in the interior of a side of a basic lattice triangle. Finally, there are $(m-1)(n-1)$ lattice points in the interior of the rectangle. The sum of the interior angles of the basic lattice triangles at each of these points is 2π because such a lattice point cannot be in the interior of a basic lattice triangle or in the interior of a side of a basic lattice triangle.

The total count is

$$4\left(\frac{\pi}{2}\right)+(2m+2n-4)\pi+(mn-m-n+1)2\pi=2mn\pi.$$

Since the sum of the interior angles of each basic lattice triangle is π, the number of basic lattice triangles is $2mn$, which is only dependent on the size of the rectangle but independent of the way it is divided. Since the total area is mn, the average area is $\frac{1}{2}$. However, since each has area at least $\frac{1}{2}$, they all have area $\frac{1}{2}$.

Here is another problem involving lattice points.

Problem 1955.3

The three vertices of a certain triangle are lattice points. There are no other lattice points on its perimeter but there is exactly one lattice point in its interior. Prove that this lattice point is the centroid of the triangle.

Let n be a positive integer. Consider all irreducible fractions with denominators at most n. If we list these fractions in ascending order, add $\frac{0}{1}$ at the beginning and $\frac{1}{1}$ at the end, we have the **Farey fractions** of order n, denoted by F_n.

For example, F_8 consists of the following.

$$\frac{0}{1},\frac{1}{8},\frac{1}{7},\frac{1}{6},\frac{1}{5},\frac{1}{4},\frac{2}{7},\frac{1}{3},\frac{3}{8},\frac{2}{5},\frac{3}{7},\frac{1}{2},$$

$$\frac{4}{7},\frac{3}{5},\frac{5}{8},\frac{2}{3},\frac{5}{7},\frac{3}{4},\frac{4}{5},\frac{5}{6},\frac{6}{7},\frac{7}{8},\frac{1}{1}.$$

If we map these fractions onto the number line, it is not easy to discern any patterns, because they are dense in some subintervals and sparse in some other subintervals. Nevertheless, there are still simple rules which govern the structure of F_n.

For the fractions $\frac{p}{q}$ and $\frac{r}{s}$, with $q > 0$ and $s > 0$, their **median** is the fraction $\frac{p+r}{q+s}$. We shall prove the following three properties of Farey fractions.

(1) The median of consecutive terms of F_n is an irreducible fraction whose denominator exceeds n.

(2) The difference between two consecutive terms of F_n is equal to the reciprocal of the product of their denominators.

(3) Each term of F_n is the median of its two neighbors.

An irreducible fraction $\frac{p}{q}$, $0 \leq p \leq q \leq n$, may be represented by the lattice point (q, p) in the coordinate plane. That $\frac{p}{q}$ is irreducible means that the line segment joining (q, p) to (0,0) contains no other lattice points. The point (q, p) is said to be **visible** from (0,0), or simply visible.

As $\frac{p}{q}$ increases, so does the angle between the positive x-axis and the line joining (0,0) to (q, p). In F_n, only (1,0) lies on the line $y = 0$, and only (1,1) lies on the line $y = x$. Moreover, the maximum value of q is n. Hence all these lattice points lie within the isosceles right triangle bounded by the lines $y = 0$, $y = x$ and $x = n$. Call this triangle H_n.

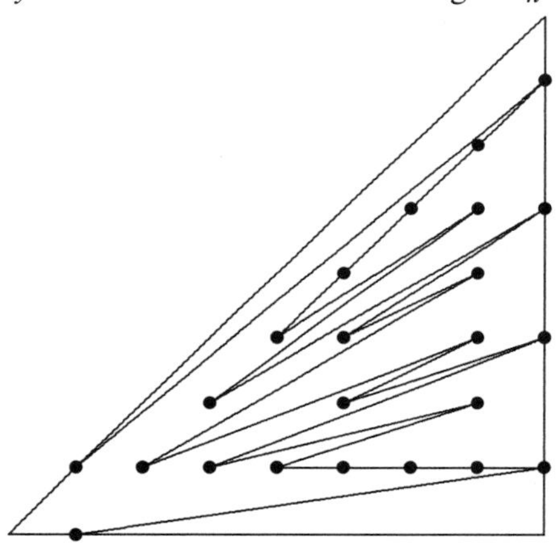

In the diagram above, we connect each pair of lattice points representing consecutive terms of F_8 by a line segment, starting from (1,0) and ending in (1,1).

Let O be the origin (0,0), A be the point (q, p) and B be the point (s, r). Then the median of the two fractions represented by A and B is represented by the lattice point C where $OACB$ is a parallelogram. With this observation, we now prove the three properties listed earlier.

(1) Let the lattice points A and B represent two consecutive terms $\frac{p}{q}$ and $\frac{r}{s}$ in F_n. Let C represent their median. Then C lies within $\angle AOB$ but

outside H_n. This is because a fraction not in F_n will have denominator greater than n. If C represents a reducible fraction, then there are other lattice points on the segment OC, and the one closest to O will lie within triangle OAB, and hence within H_n. This is a contradiction.

(2) From (1), we see that OAB is a basic lattice triangle. That A and B are visible means that there are no other lattice points on OA or OB. Since OAB lies with H_n, there are no lattice points in OAB or on AB.

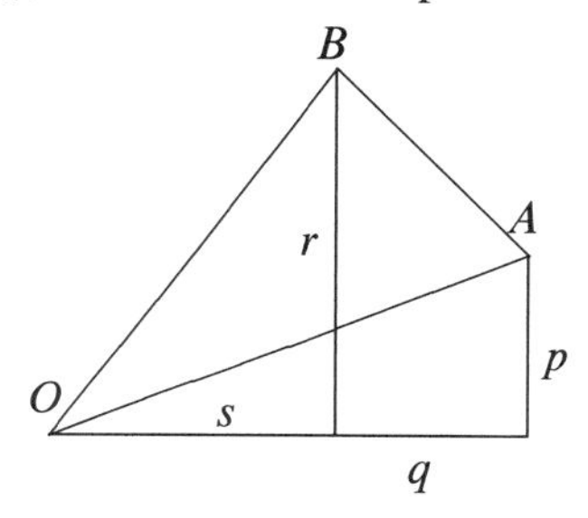

As was proved in Problem 1942.2, the area of a basic lattice triangle is $\frac{1}{2}$. Now the area of OAB is

$$\begin{aligned}\frac{1}{2} &= \frac{1}{2}rs + \frac{1}{2}(q-s)(p+r) - \frac{1}{2}pq \\ &= \frac{1}{2}(qr - ps).\end{aligned}$$

Hence $\frac{r}{s} - \frac{p}{q} = \frac{qr-ps}{pq} = \frac{1}{pq}$.

(3) Let A, B and C represent three consecutive terms of F_n. Then OAB and OBC are basic lattice triangles. Hence A and C lie on opposite sides of OB and equidistant from OB. It follows that OB passes through the point D where $OADC$ is a parallelogram. Now D represents the median of $\frac{p}{q}$ and $\frac{r}{s}$. Hence so does B and we must have $D = B$.

The geometric interpretation of (3) is that if a lattice triangle contains lattice points other than its vertices and they lie on a line through a vertex, this line must be a median of the triangle. If a lattice triangle contains exactly one lattice point in its interior, then this point lies on all three medians, and must be the centroid of the triangle. This furnishes an alternative solution to Problem 1955.3.

The inner product of vectors, treated in Hungarian Problem Book III, can be used to provide simple solutions to the following two problems from Hungarian Problem Book II.

Problem 1912.3

Prove that the diagonals of a quadrilateral are perpendicular if and only if the sum of the squares of one pair of opposite sides equals that of the other.

Let the quadrilateral be $ABCD$ and let O be an arbitrary point. Then

$$\begin{aligned}
AB^2 + CD^2 - BC^2 - DA^2 \\
&= (\mathbf{OB} - \mathbf{OA})^2 + (\mathbf{OD} - \mathbf{OC})^2 - (\mathbf{OC} - \mathbf{OB})^2 - (\mathbf{OA} - \mathbf{OD})^2 \\
&= \mathbf{OA}^2 + \mathbf{OB}^2 - 2\mathbf{OA} \cdot \mathbf{OB} + \mathbf{OC}^2 + \mathbf{OD}^2 - 2\mathbf{OC} \cdot \mathbf{OD} \\
&\quad - \mathbf{OB}^2 - \mathbf{OC}^2 + 2\mathbf{OB} \cdot \mathbf{OC} - \mathbf{OD}^2 - \mathbf{OA}^2 + 2\mathbf{OD} \cdot \mathbf{OA} \\
&= 2\mathbf{OB} \cdot (\mathbf{OC} - \mathbf{OA}) + 2\mathbf{OD} \cdot (\mathbf{OA} - \mathbf{OC}) \\
&= 2(\mathbf{OB} - \mathbf{OD}) \cdot (\mathbf{OC} - \mathbf{OA}) \\
&= 2\mathbf{DB} \cdot \mathbf{AC}.
\end{aligned}$$

Both conditions in the problem are satisfied if and only if the displayed expression has value 0.

Problem 1918.1

Let AC be the longer diagonal of the parallelogram $ABCD$. Drop perpendiculars from C to AB and AD extended. If E and F are the feet of perpendiculars, prove that $AB \cdot AE + AD \cdot AF = AC^2$.

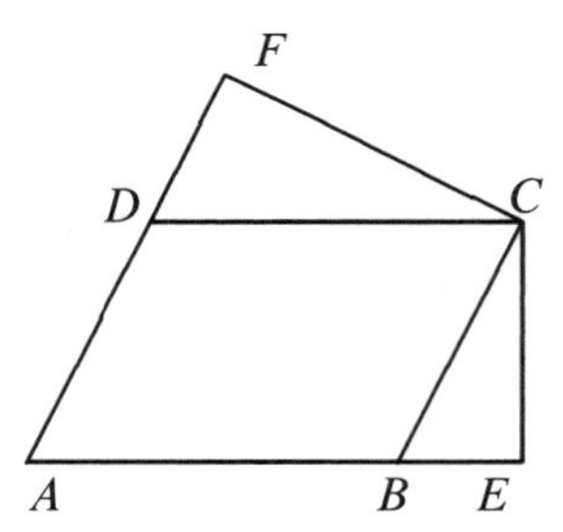

Since the vectors **AB** and **EC** are perpendicular, we have

$$AB \cdot AE = \mathbf{AB} \cdot \mathbf{AE} = \mathbf{AB} \cdot (\mathbf{AC} - \mathbf{EC}) = \mathbf{AB} \cdot \mathbf{AC}.$$

Similarly, we also have $AD \cdot AF = \mathbf{AD} \cdot \mathbf{AC}$. It follows that

$$AB \cdot AE + AD \cdot AF = (\mathbf{AB} + \mathbf{AD}) \cdot \mathbf{AC} = \mathbf{AC}^2 = AC^2.$$

The following problem, also from Hungarian Problem Book II, contains a surprise.

Problem 1924.2

If O is a given point, ℓ a given line and a a given positive number, find the locus of points P for which the sum of the distance from P to O and from P to ℓ is a.

The answers to locus problems are usually parts of straight lines and circles. Here, it is parts of two intersecting parabolas! Nowadays, parabolas are either not studied at all, or treated as some algebraic equations like $y = x^2$ in analytic geometry. However, to solve Problem 1924.2, it is necessary to know the geometric definition of a parabola.

A **parabola** is the locus of points which are equidistant from a given line and a given point not on that line. The given point is called the **focus** and the given line the **directrix** of the parabola. The line through the focus and perpendicular to the directrix is called the **axis** of the parabola, and the parabola is symmetric about this line. The point of intersection of the axis and the parabola is called the **vertex** of the parabola.

Let the distance between the focus and the directrix of a parabola be $\frac{1}{2}$. Set up a coordinate system as follows. Let the focus be the point $(0, \frac{1}{4})$ and the directrix be the line $y = -\frac{1}{4}$. Then the axis is the line $x = 0$ and the vertex is the point (0,0). Let (x, y) be a point on the parabola. Its distance from the focus is $\sqrt{x^2 + (y - \frac{1}{4})^2}$ and its distance from the directrix is $y + \frac{1}{4}$. Equating and squaring yields $x^2 + y^2 - \frac{y}{2} + \frac{1}{16} = y^2 + \frac{y}{2} + \frac{1}{16}$, which simplifies to the familiar $y = x^2$.

The following problem deals with half-planes, which are special cases of convex figures.

Problem 1951.3

A plane can be covered completely by four half-planes. Prove that three of these four half-planes are sufficient for covering the plane completely.

A plane figure is said to be **convex** if for any two points in the figure, the entire segment joining the two points lies entirely within the figure. Examples of convex figures are circles, triangles, half planes, infinite strips, regions between two arms of angles, straight lines, rays and segments.

From the definition, it can be deduced that intersections of convex figures are also convex. The intersection of all convex figures containing a figure F is called the **convex hull** of F. The convex hull of a finite point set is either a segment if the points are collinear, or a convex polygon otherwise.

The following is a basic result about convex figures.

Helly's Theorem *Among $n \geq 4$ convex figures, if any three has a common point, then they all have a common point.*

Proof We use mathematical induction on n. We first consider the base case $n = 4$. Let F_1, F_2, F_3 and F_4 be convex figures such that F_2, F_3 and F_4 have a common point P_1. Similarly, there is a point P_2 which is in F_3, F_4 and F_1, a point P_3 in F_4, F_1 and F_2 and a point P_4 in F_1, F_2 and F_3.

The convex hull of P_1, P_2, P_3 and P_4 is either a segment, a triangle or a convex quadrilateral. Suppose it is a segment, say P_1P_4. Then P_2 is on the segment. Since both P_1 and P_4 belong to F_2, so does P_2, and it is a common point of all four convex figures.

Suppose the convex hull is a triangle, say $P_1P_2P_3$. Let the extension of P_1P_4 cut P_2P_3 at some point P. Then P, being on P_2P_3, belongs to F_4. Hence P_4 also belongs to F_4. Finally, suppose the convex hull is a convex quadrilateral, say $P_1P_2P_3P_4$. Let the P be the point of intersection of P_1P_3 and P_2P_4. It is not hard to show that P belongs to all four convex figures.

Suppose the result holds for some $n \geq 4$. Consider the next case with figures F_1, F_2, $\ldots$, F_{n+1}. Define F to be the intersection of F_n and F_{n+1} so that we are back to n convex figures.

We claim that every three of them have a common point. This is certainly true unless one of the three figures is F. For a triple F_i, F_j and F, $1 \leq i < j \leq n-1$, replace F by F_n and F_{n+1}. For these four convex sets, it is given that every three have a common point. Hence all four have a common point by the base case. This point will then be a common point of F_i, F_j and F, justifying the claim.

We can now apply the induction hypothesis to show that F_1, F_2, $\ldots$ F_{n-1} and F have a common point. This point is also the common point of all $n+1$ convex figures. This completes the proof of Helly's Theorem.

We can now give an alternative solution to Problem 1951.3. For each of the given half-planes, consider its complementary half-planes. If every three of the complementary half-planes have a common point, then all four complementary half-planes have a common point by Helly's Theorem. However, since the four given half-planes cover the entire plane, the four complementary half-planes cannot have a common point. This means that some three of the complementary half-planes have no common points. In other words, the corresponding three given half-planes already cover the entire plane.

About the Editors

Robert Barrington Leigh (1986–2006) was one of the very best students in thirty years of Andy Liu's Saturday Mathematical Activities, Recreations & Tutorials program. When in Grade 6, he coauthored a paper, with a friend in Grade 7, on one of the problems in this book. It was first published in Australia, and then translated and published in Hungary. The two teamed up for another paper in the MAA's *College Mathematics Journal* the following year. Robert won two Bronze Medals at the International Mathematical Olympiad on his pure talent, without doing extra training. The same approach earned him seventh to sixteenth place in the Putnam all three times he participated in the Competition. Tragically, he died before he could enter the Competition for the fourth time.

Andy Liu received a BS degree with First Class Honors in Mathematics from McGill University in 1970. He earned his MS in Number Theory (1972) and a Doctor of Philosophy in Combinatorics (1976) from the University of Alberta, Edmonton. Dr. Liu won a Minnesota Mining & Manufacturing Teaching Fellowship in 1998. He has won numerous teaching awards including: Canadian University Professor of the Year in 1998 (awarded by the Canadian Council for the Advancement of Education and the Council for the Advancement and Support of Education); the Canadian Mathematical Society's Adrien Pouliot Education Award in 2003; and the Mathematical Association of America's Deborah and Franklin Tepper Haimo Teaching Award in 2004. Dr. Liu is very well known in problem-solving circles. He won the David Hilbert International Award in 1996 from the World Federation of National Mathematics Competitions. He was the Deputy Leader (under Murray Klamkin) of the USA Mathematical Olympiad Team from 1981–1984; he later was Leader of the Canadian IMO team. Dr. Liu has been the Editor of the problem section of the MAA's *Math Horizons* for seven years.